ASTRONOMY
in colour

by Peter Lancaster Brown

BLANDFORD PRESS
POOLE   DORSET

First published 1972
Revised edition 1973
Revised edition 1975
Reprinted 1979
Reprinted 1983

ISBN 0 7137 0729 1

Distributed in the United States by
Sterling Publishing Co., Inc.,
2 Park Avenue, New York, N.Y. 10016.

To my loving wife

Printed and bound in Great Britain by
Richard Clay Ltd.,
Bungay and Norwich

PREFACE

Inevitably it has become a scientific cliché to repeat that astronomy and space research is the fastest developing of all sciences, yet this is an inescapable fact. Almost every week science magazines announce new discoveries in space and provide the latest fashionable opinion about this or that theory of the Universe. As a result no book on astronomy can be entirely up to date after going to press.

Since the last edition of this book was printed, there have been some interesting discoveries and some unexpected ones. For example, it is now known that Uranus and Jupiter are ringed planets like their neighbour Saturn. A new planet, albeit a small one, has also been discovered wandering between the orbits of Saturn and Uranus. This new planet has been approximately named Chiron by astronomers because Chiron – a centaur – was a descendant of the gods Saturn and Uranus. It is now thought that Chiron probably represents one of a number of as yet undiscovered dimunitive planetary-like bodies wandering in the more distant regions of the solar system.

A great deal more is now also known about the surfaces of the nearer planets. Huge volcanoes and mini-planet impact craters are seen to dominate the harsh, rugged landscapes of Mars, Venus and Mercury. In its earlier days it was likely that the Earth had a similar appearance, but the complex geological processes at work over several aeons have all but removed the earlier scars. The Viking missions to Mars have provided tantalising evidence that life *might* be possible there, but extant life itself on the planet's surface, so far explored, is still not proven.

Most of the new discoveries in the solar system are the net result of space probes such as the Viking missions to Mars (above) and the grand tour Pioneer and Voyager missions to the outer planets (*see* Fig. 76). When Voyagers 1 and 2 reached Jupiter in 1979, they identified features that went far beyond anything that could be seen by Earth-bound observers. For the first time positive evidence about the surfaces and nature of Jupiter's largest satellites (*see* p. 80) came to hand. Callisto was found to be a giant ball of ice with a heavily cratered surface. Ganymede was seen to have fewer craters, but there were some very strange grooves traversing its surface which were likely formed when its crust cooled and contracted to leave a distinctive ice-wrinkled pattern. Few craters are directly visible on this moon because the ice crust has oozed up from below and obscured them.

However, the weirdest of Jupiter's moons is Europa which has the appearance of a cracked, white billiard ball. To date only three impact craters have been seen on its surface, and it is concluded that the surface of Europa is covered in a mantle of thick slushy ice that soon obliterates any new craters formed there.

One surprise resulting from the Voyager missions came from the fourth, largest satellite Io. Before space missions reached Jupiter, it was expected that this moon *would* be heavily cratered. However, not a single impact crater has so far been identified. Nevertheless, Io has turned out to be the most surprising moon of all, for no less than eight active volcanoes have been detected on its surface. It is believed that at present Io is the most active volcanic body in the entire solar system.

Among other Voyager discoveries was the detection of a new, fourteenth moon and several rings around Jupiter. It was also found that great lightning storms perpetually rage in the planet's atmospheric belts.

The Pioneer II probe (*see* p. 106) encountered Saturn in late August/early September 1979. One major discovery was that Saturn has a magnetic field which is unique in that its axis corresponds almost exactly with the planet's axial rotation. In contrast, the magnetic fields of the Earth, Jupiter, Mercury and the Sun are all tilted at about 10°.

The Pioneer spacecraft detected radiation belts round Saturn like those known round Jupiter and the Earth (*see* Fig. 37), but Saturn's radiation belts are also unique because the rings periodically 'mop up' all charged particles, resulting in a situation where the region inward of the outermost edge of the ring system is probably the most radiation-free zone in the entire solar system.

Pioneer II also detected a new satellite of Saturn (designated S-2), but it provided no confirmation of the elusive satellite Janus (*see* p. 83). In 1980 and 1981, the Voyagers passed close to Saturn and provided the most spectacular views of the rings to date, revealing them to be highly complex structures with many more divisions than seen from Earth. The latest observations of Saturn's rings has swung opinion back to an earlier theory that they consist of icy particles (*see* p. 82). These particles are certainly several centimetres in size, and the rings themselves consist of more than one layer of material. At the time of writing, Voyager 2 is heading towards its rendezvous with Uranus in 1986 and then hopefully it will reach Neptune in 1989.

Beyond the solar system, in deep cosmic space, Black Holes now seem to be a reality rather than just a bizarre theory. Observations of the very distant so-called radio galaxies have revealed that their component parts are separating at speeds several times the velocity of light. Some say: Impossible! Maybe, but explanations how these superluminal velocities are to be reconciled within the established laws of physics perplex and torment the leading cosmologists of our day.

Peter Lancaster-Brown
February 1983

CONTENTS

Phobos, one of the tiny satellites of Mars photographed by Mariner 9. Phobos revolves round Mars in 7 hrs 39 mins, and its maximum diameter is 22 km (13·7 miles) (see p. 73).

ACKNOWLEDGEMENTS

The author and publisher gratefully acknowledge the following sources for illustrations and photographs other than the author's:

Mr. John Wood (34, 36–7, 42–3, 45, 53, 75–7, 84, 86–7, 89–90, 110–12, 125); Mount Wilson and Palomar Observatories (60a, 63a, 64a, 65, 85, 91, 94, 97–109, 113, 118–19, 124, 126–27, 140, 153–54, 159); National Aeronautics and Space Administration (35, 48–52); United States Naval Observatory (38c, 63b, 64b, 69, 88, 92–3, 95–6, 109, 114–17, 120–23); Lick Observatory, University of California (62, 130, 142, 158); Royal Astronomical Society (141, 143, 145–46, 155–57, 162b); Ronan Picture Library (163); Lunar and Planetary Laboratory, Tucson (60b, 161); American Meteorite Laboratory (71–4, 149); Mr. W. Baxter (46); Australian News and Information Bureau (128–29, 148, 160, 164–65); Lowell Observatory (144); Novosti Agencies (162); Mr. H. Dall (47); Mr. C. Hunt (39, 55–9, 139, 152); British Museum (29–30, 33, 83, 147); Diana Wyllie Ltd (38a/b, 40).

The author is especially grateful to the illustrator John W. Wood for his realistic and accurate colour interpretations and to the staff of Blandford Press for their general enthusiasm and helpfulness in all matters. Lastly, but by no means least, he wishes to acknowledge the generous and unflagging assistance from his long-suffering wife, Johanne, who spared the author much of the routine drudgery.

6

INTRODUCTION

Although the development of modern astronomy owes much to the progress of other natural sciences such as physics, chemistry and geology, it was the simple movement of the night sky which first kindled the spark of natural curiosity in ancient man.

Astronomy has advanced as a science through different ways. The advances were the results of philosophical speculation, new discoveries, instrumental invention or progress in other kindred scientific fields.

The characteristics of a big science is that it should have a marked influence on other sciences. Astronomy as a big science is the dominant parent to the science of astrophysics which depends on astronomy and physics for its intellectual nourishment and progress, for new ideas need to be tried out in both the environments of the terrestrial laboratory and in the depths of cosmic space.

In prehistoric times, after the invention of farming, astronomy played a fundamental role in regulating the patterns of everyday life. Through the discovery of the harmonic shifts of the star sphere man was able to gauge the passage of time and so anticipate the onset of the various seasons when crops were to be sown and reaped. The great stone circles of megalithic north-west Europe – built before the pyramids and once thought of as rude edifices to forgotten pagan gods – are now recognized as scientific tools of a sophisticated race of ancient men who designed and erected them to serve as astronomical observatories.

The Greeks laid the true foundation stone of modern astronomy when they synthesized the knowledge of the ancient world and formulated the basic concepts about the Earth and Sky. During the Dark Ages this knowledge was retained and added to by the Arab nations until the time of the Renaissance.

The establishment of the heliocentric nature of the Universe, followed soon after by the invention of the telescope and the publication of Newton's *Principia*, made astronomy the most influential science of the sixteenth, seventeenth and eighteenth centuries. Not until the invention of the spectroscope, the understanding of the electromagnetic nature of light and the development of modern theories about the atomic nature of matter did fundamental thinking about the nature of the Universe begin to change. The last great dawning began with the far-reaching space-time ideas of Einstein and the construction of giant optical and radio-telescopes which could penetrate the depths of cosmic space and provide evidence about the distance of the visible Universe. Today with the discovery of the enigmatic pulsars, quasars, the bizarre anti-matter and the ability to examine bodies in the solar system by *in situ* observations using space probes, many believe that astronomy will soon take another exciting leap forward.

Certainly astronomy is no longer the obscure science of a few decades ago pursued only by those in the rarified atmosphere of academic establishments. The advent of the Space Age has brought it back into the reality of the everyday world as much as it once was with our distant forebears who knew the stars and constellations as well as their more mundane earthly surroundings. In the present age almost everyone has pondered over questions such as: Is there intelligent life elsewhere in the Universe? Can man ever reach the stars? Is the Universe finite? and What 'star' is that? when an artificial satellite is spotted stealing across the night sky.

One of the prime motives of science has been the pursuit of knowledge about the structure and nature of matter. Perhaps one of the greatest attractions of astronomy is that it is the science most immediately identified with the problems of cosmology: the understanding of the everyday world and the Universe in its entirety. All science is, of course, cosmology, but it is astronomy with its preoccupation with a whole range of phenomena extending from the microscopic to the macroscopic which provides the test-bed for cosmological ideas.

I THE EARTH IN SPACE

The Dawn of Astronomy

There can be no doubt that all primitive races were greatly interested in the heavenly bodies from the earliest times. It was not, however, until the beginning of agriculture in c. 10000 B.C. that man began to notice and systematically record their rhythmical patterns. The sky played an important part of everyday living. In nearly all primitive societies the Earth and the sky evolved as a pair of gods or goddesses long before there was any real understanding of astronomy as a science.

In ancient Egypt the sky was Nu and was represented as a female figure bending over Seb, the Earth, with her feet on one horizon and her fingertips on the other (Fig. 29). Nu and Seb were separated by Shu, the god of air or sunlight. The Sun's daily journey was represented by the god Ra seated in a boat traversing the sky from east to west.

The Egyptians conceived the Earth as a flat platter with a corrugated rim. This platter floated on the water. The waters below the platter were termed the abysmal regions and given the name Nun, the waters of the underworld. It was Nun who was reckoned to be the primordial source of life. The sky took on the appearance of a roof or celestial vault on which all the heavenly bodies were represented as lights situated at the same distance, side by side. The sky revealed itself as a remote awesome presence to both the Egyptians and the Babylonians. It suggested an omniscient power so great that it commanded allegiance simply by its appearance. We can read from the ancient texts that it created a kind of 'welling-up' of man's soul and brought about the realization of his insignificance. It is of interest that man today is still receptive to this mood, in spite of his more materialistic approach to life.

The earliest scientific astronomies arose almost simultaneously in Egypt, Babylonia and in north-west Megalithic Europe. The agricultural Egyptian and Babylonian civilizations ran parallel chronologically until the time of the Persian Empire c. 500 B.C. In both cultures we meet the same, and also some highly contrasting, cosmological ideas. Water was very significant as a source and the sustenance of life. The Egyptians made practical use of the stars to signify the changing of the seasons. The heliacal rising of the bright star Sirius (α Canis Majoris) was used to herald the oncoming of the annual Nile flood and conveniently used to denote the beginning of the new year. Unfortunately, such scientific predictions – actually the forerunner of the modern science of meteorology – became interwoven with less scientific mythological ideas. Astrology probably arose when men, after successfully predicting the seasons by star risings, thought it reasonable to extend their predictions to the fate of human beings.

At this early age then, astrology and astronomy could not be separated in either of the Middle Eastern cultures. In Egypt, we can read in the Pyramid texts that the goal of the deceased was the region of Dat, in the northern part of the heavens. He who joined the circumpolar constellations (which are always visible throughout each night of the year) would live for ever. Nevertheless, from Egyptian astronomy we derive our modern calendar, and from Babylonian astronomy we derive the sexagesimal system (of minutes in the hour) based on the number 60.

After the invention of writing in c. 3000 B.C., we can learn a great deal about early Middle Eastern astronomies from hieroglyphic or cuneiform texts. However, in north-west Europe, where the astronomy of the Megalithic culture flourished, we can glean no such information since writing, apart from a very primitive picture form, was quite unknown. Yet, the great stone monuments, left by the Megalithic races, tell their story indirectly. In many instances the megaliths (from the Greek *megas*, 'great' and *lithos*, 'stone') are arranged in circles or sometimes in an egg-shaped pattern. Formerly these circles were considered to be pagan temple sites built specifically for ritual or sacrifice. However, present-day investigations have shown quite conclusively that they were astronomical observatories laid out to a precision which matches that achieved by the modern surveyor using optical equipment.

Avebury Circle, England (Fig. 30), is a fine example of Megalithic astronomical architecture, laid out to an accuracy of 1:1000, using the basic linear unit of a Megalithic yard 83 cm (or 2·72 ft). Stonehenge is another such example which was probably constructed to predict lunar and solar eclipses that presumably regulated the contemporary Megalithic calendar. Recent Carbon-14 dating, which utilizes the principle of radioactive isotope decay to give accurate chronological ages, now dates the Megalithic astronomical culture *before* the Egyptian. This new fact is a great surprise to those concerned with the history of astronomy since it was always previously supposed that any astronomical knowledge came to north-west Europe from Egypt and the Middle East by the process of diffusion. The Megalithic culture appears to have been highly advanced, and from the arrangement of the stone circles it can be inferred that their astronomers knew the properties of the right-angled triangle long before the Greek, Pythagoras, formulated it mathematically. But Megalithic astronomy was dead long before the Roman conquests, and its sudden decline remains a mystery.

Greek Astronomy

The Hellenic period of science provided a melting-pot for all the various astronomies inherited from the ancient world; yet the chief influences were the Babylonian and Egyptian ones which formed the foundation stone of all astronomical belief handed on and modified by the Greeks right up to the Renaissance period.

Oriental astronomies such as those derived from China and Japan had little or no bearing on Greek ideas. It is only in the nineteenth and twentieth centuries that modern astronomy has made use of their thoroughly documented and sophisticated records of comets and 'guest stars' to fill the long gaps in observational astronomies of the earlier western world. Neither did the Greeks have any knowledge of the Megalithic astronomies, which were not revealed until very recently. When we think of Greek astronomical ideas, we must remember that the majority of manuscripts on which our knowledge of Greek science is based are contained in the 'second-hand' sources of the Byzantine codices, written 500 to 1500 years after the lifetimes of their Greek authors.

The Greeks inherited the idea of the spherical nature of the Universe from the older civilizations, but by the sixth century B.C., they had begun to suspect that the Earth itself took the form of a globe rather than a flat dish. The Greek philosopher Anaximenes (585–528 B.C.) is attributed with the first observational proof of its spherical nature when he noted the circular shadow cast on the Moon at the time of a lunar eclipse. Likewise Pythagoras, in 550 B.C., taught his pupils that the Earth was a globe floating freely in space. However, it can be fairly certain from more fragmentary references that long before this period man had noticed the now commonplace proof of the Earth's sphericity by the disappearance of ships below the extended sea horizon.

Eratosthenes (276–194 B.C.), the fifth librarian of the Alexandria Library, evolved another ingenious experiment to test the theory and obtain an accurate estimate about the Earth's size.* As librarian, Eratosthenes gleaned some interesting information. In a deep well at Syene (the site of the modern Aswan Dam) it had been noted that the Sun's rays fell vertically down at midsummer, while at Alexandria, almost due north of Syene, they were $7\frac{1}{2}°$ from the vertical. From the records supplied by the tax collectors, he knew the linear distance between the two geographical points, and from his knowledge of mathematics he was able to make the necessary calculation. His result was a circumference of 40,250 km (25,000 miles) (equivalent to 250,000 stadia, the contemporary Egyptian units), a surprisingly close value

* Before this only inspired guesses were made. Aristotle quoted 64,400 km (40,000 miles).

11

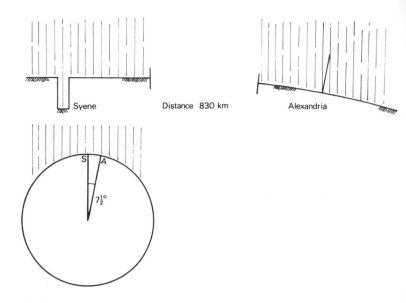

Fig. 1 Eratosthenes' method of measuring the size of the Earth. At Syene, in midsummer, the Sun's rays are vertical, at Alexandria they make an angle of $7\frac{1}{2}°$. Knowing the distance to be $\simeq 830$ km, Eratosthenes estimated the Earth's circumference to be $\simeq 830(360 \div 7\frac{1}{2})$ km.

to that accepted today. It is often quoted that Eratosthenes conducted the experiment himself at Syene and at Alexandria, but this seems very doubtful, for he was by all accounts principally a literary man, and it would appear he simply made excellent opportunity of his access to information.

Aristarchus of Samos (310–230 B.C.) invented a way of measuring the distance of the Sun using the method of right-angled triangles. Although his final result was wide of the mark, the method was correct. He also put forward the idea of the heliocentric nature of the solar system.

Since the earliest times, the daily movements of the Sun, Moon and the stars, and the yearly movements of the planets against the backcloth of stars had been a puzzling phenomenon. In the fourth century B.C., Heracleides of Pontus first suggested that it was the Earth itself which turned on its axis rather than the sky moving round the Earth. Some of Heracleides' other theories were not quite so sound – including his idea that the Sun's diameter measured only a foot!

Aristarchus' suggestion that the Earth might travel round the Sun failed to gain permanent favour. An alternative theory put forward by Ptolemy (A.D. 120–180) which reverted to earlier Babylonian ideas, took root instead. There can be no doubt that Ptolemy was the most influential astronomer of his day. His system of circles (Fig. 2) seemed to be the best compromise in explaining the peculiarities of planetary motion, and it became universally adopted. Ptolemy considered that the Earth lay at the centre of the Universe, and round it revolved the Moon, the Sun, the planets and the entire visible sphere of fixed stars in paths which described perfect circles. At this period Greek astronomy was obsessed with the circle and sphere, which had been adopted as the perfect forms. It became an essential part of their philosophy that nature should fit the pattern of perfection.

Fig. 2 The Ptolemaic system (not to scale). According to Ptolemy the Sun and the planets moved round the Earth in circular paths or *deferents*. In addition the five planets were also considered to move in circular orbits called *epicycles*. Mercury and Venus were supposed always to be positioned in a straight line between the Earth and the Sun. In the case of the three outer planets the line joining the epicycle to the centre was supposed to remain parallel to the Earth–Sun line.

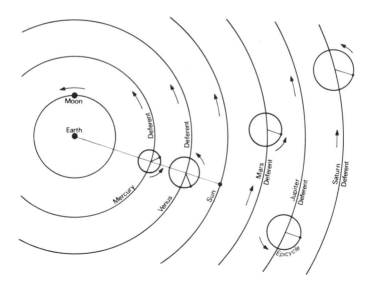

Uncommitted observers, however, soon detected anomalies of motion which would not fit the idealized system. Ptolemy was required to introduce a whole complexity of additional motions called epicycles and deferents (or circles round circles, *see* Fig. 2) in order to satisfy his more pragmatic contemporaries.

Renaissance Astronomy

Because of its theoretical perfection, the Ptolemaic planetary system was officially adopted by the early Church. Doubtless, owing to the primary importance of the Earth's position in this system, it left them no alternative. Despite the overwhelming observational objections, the idea persisted and was maintained right up to the Renaissance in Ptolemy's great literary work the *Almagest*, which had become the 'scientific bible' over a period exceeding 1200 years.

But in the sixteenth century, the awaking began. In 1543, Nicolaus Copernicus published his monumental book *De Revolutionibus Orbium Coelestium*, which restated the earlier Greek heliocentric ideas that placed the Earth in space as a satellite planet to the Sun together with all the other planets belonging to the solar system. Yet the way was still not clear for a universal change from the old views that had held sway for over 1500 years. Some sixteenth-century observers, including Tycho Brahe, the last great pre-telescopic observer, objected that Copernicus had not fully solved the problem. Indeed, Tycho Brahe maintained to his death in 1601 that the geocentric system was fundamentally the correct one in spite of its drawbacks. Ironically, it was Tycho's own painstaking observations of Mars, inherited by his pupil Johannes Kepler, which finally provided an astronomical 'Rosetta stone' to unravel the complexities of planetary motion which had bothered scientific humanity for so long.

The Earth's Orbit

After Copernicus put forward the idea of the Sun as the central body of the solar system, some of the objectors to his theory rightly pointed out that if it were correct, surely a revolving Earth ought to cause a displacement of the stars during the course of its motion. But even Copernicus had no concept of the vast linear distances that stretched between the Earth and the nearest stars. It was not until 1838 that the first displacements were measured and the distances calculated. However, long before Copernicus' time, it had been realized that the old pre-classical idea of the physical rotation of the entire star sphere around the Sun was extremely unlikely.

When Kepler (1571–1630), and later Newton, investigated the problem of planetary motion, it was found that the Earth does not revolve round

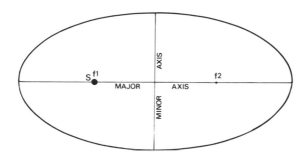

Fig. 3 The properties of an ellipse. The eccentricity is measured by expressing the distance between the two foci, f¹ (Sun) and f², in terms of the length of the major axis.

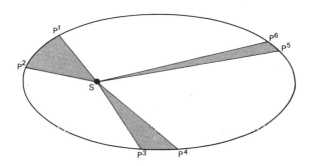

Fig. 4 Kepler's second law (equal areas). A planet (or comet) takes the same time in moving from P¹–P², P³–P⁴, and P⁵–P⁶. The swept area P¹–S–P²=P³–S–P⁴=P⁵–S–P⁶.

the Sun in a perfect circle, so beloved by the Greeks. Instead its path describes very nearly the figure of an ellipse. Kepler in using the observations of Mars, inherited from Tycho Brahe, formulated three basic laws of planetary motion, published in the *Rudolphine Tables* (after the name of his patron, Emperor Rudolf II), which stated:

(1) The orbit of each planet is an ellipse having the Sun in one of its foci.
(2) The motion of each planet in its orbit is such that the radius vector from the Sun to the planet describes equal areas in equal times.
(3) The squares of the periods in which the planets describe their orbits are proportional to the cubes of their mean distances from the Sun.

Later, Isaac Newton (1642–1727) continued where Kepler left off and proceeded to demonstrate that gravity was the dominant factor affecting the motion of all bodies. In his classical book, *Principia*, Newton shows that:

(1) A body remains at rest or continues to move in the same straight line with constant speed unless it is acted upon by another force.
(2) The force applied to a body is in the direction of the acceleration imparted to the body, and is equal to the mass of the body times its acceleration.
(3) Every action has an equal and opposite reaction.

Newton's work was the greatest step forward astronomy has ever known. It has been said that one can perfectly well understand the *Principia* of Newton without much knowledge of earlier astronomy, but that one cannot read a single chapter in Copernicus, or understand Kepler, without a thorough knowledge of Ptolemy's *Almagest*. The reason is of course that Newton, with his introduction of dynamical mathematics, broke completely new ground, as did Einstein in more modern times. Before Newton, all astronomy consisted in modifications of Egyptian, Babylonian and later Hellenistic ideas.

The property of an ellipse provides for the Sun being located at one focus, and the degree of eccentricity expresses the shape of an ellipse which is defined by the ratio of the distance between the two foci to the longest diameter (Fig. 3). All ellipses have eccentricities expressed between zero and one. The eccentricity of a circle is zero.

Because of the Earth's elliptical-shaped orbit, we are nearer to the Sun at the time of mid-winter in the northern hemisphere. This point is referred to as the Earth's perihelion (derived from the Greek *peri*, 'near' and *helios*, 'sun'). The furthest distance, at the time of mid-summer in the northern hemisphere, is termed aphelion (Greek *ap*, 'away').

Another property of the ellipse is variable orbital speed, as shown in Kepler's Second Law (*see* Fig. 4), so that when a planet is nearer the Sun, it is moving faster than when at aphelion. However, in the case of the Earth's orbit, the eccentricity is very small (0·01674), and the ratio between the two axes is less than 1 : 1000. This means that the Earth completes one half of its revolution round the Sun in the period spring to autumn (northern hemisphere) in 186 days and from autumn to winter (northern hemisphere) in 179 days.

The movement owing to the Earth's motion round the Sun can be noticed in the sky by night-to-night observation. This movement has the effect of bringing a particular star or constellation to the same position in the sky 4 minutes *earlier* each night. If one multiplies the 4-minute interval per night by 365, this equals approximately 24 hours and therefore also equals one total revolution of the sky, and thus after one year brings the Earth back to the starting point again.

The Earth's Rotation

In the earlier astronomies we see that the star sphere was supposedly turning about the Earth. Apart from the evidence of the daily movement of the stars, Sun and Moon, the simplest way to detect the Earth's rotation is to observe the behaviour of a pendulum. This experiment was first performed by the Frenchman Foucault in 1851. He suspended a massive pendulum ball, about 30·5 cm (1 ft) in diameter, by means of a 61 m (200 ft) long wire from the dome of the Panthéon in Paris.

If a pin or spike is set into the base of the ball so that it can trace out a pattern in a layer of loose sand, it will soon be seen, after a number of back and forth oscillations, that the pendulum actually appears to rotate slowly to the right (from east to west), which is the direction *opposite* to that of the rotation of the Earth. If the experiment is conducted in the southern hemisphere, the oscillations take place from west to east.

The pendulum experiment is also a demonstration of Newton's First Law, for the pendulum always swings in one direction unless a force acts on it at right angles to the plane of oscillation (Fig. 5). Of course, there

Fig. 5 Pendulum experiment. Schematic drawing showing the Earth rotating while a pendulum plumb bob continues to swing in a fixed plane.

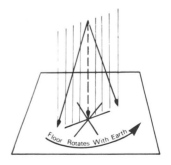

is no force acting on it, so that we conclude it is the surface of the Earth turning under the swing of the pendulum.

If the experiment is repeated at the poles, the pendulum will rotate once in 24 hours. Between the poles the period of revolution takes longer and is dependent on the latitude where the experiment takes place. For example, in New York the time taken would be about 36 hours. One calculates the period of time for any particular latitude from the simple formula: 24/sin λ, where λ equals the latitude.

Time

The rotation of the Earth also provides us with a basis for time, but it will be seen that a more accurate basis is the Earth's revolution round the Sun. The 24-hour system is another by-product of early Egyptian astronomy. During the period of the Middle Kingdom, 12-hour star clocks were often portrayed on coffin lids and on the ceilings of royal tombs (Fig. 32). Egyptian time reckoning was a decimal system consisting of 10 hours during daylight and a duodecimal system at night consisting of 12 hours. The remaining 2 hours were reckoned for twilight periods. Later the Greeks rationalized the 24-hour clock into the one we use today, which included the Babylonian sexagesimal number system so that each hour was subdivided into 60 minutes.

If we gauge the rotation of the Earth from the star sphere, it completes one revolution in 23 hrs 56 mins. This is called a sidereal day, or star day, because sidereal means star. It may be remembered, however, that the Earth in its orbit round the Sun takes 365¼ days to complete one revolution. This results in the Sun appearing to lag 4 minutes behind each day, hence solar time, i.e. time measured by the Sun, is *longer* than sidereal time.

Nevertheless, daily life is regulated by solar time and *not* by sidereal time with which astronomers are more concerned. But, unfortunately, solar time is not of continuous length. The inequality arises from two factors:

(1) The orbit of the Earth round the Sun is not circular, causing a seasonal difference in the speed at which the Earth travels.
(2) The Sun moves in the path of ecliptic where it is inclined to the celestial equator at an angle of 23½°.

To overcome the unequal seasonal length of day, the term *mean solar time* is used, which is based on a fictitious Sun that moves round the sky at a uniform rate. This provides for days of equal length and is important in civil time.

The non-uniform motion of the Sun gives rise to the *Equation of Time* –

or simply the difference one can see if one compares mean solar time (clock time) with sun-dial time. The maximum possible difference amounts to approximately 16 minutes and occurs in November (*see* table).

THE EQUATION OF TIME

(Apparent Time minus Mean Time)

Date		Minutes	Date		Minutes
January	1	— 3·2	July	1	— 3·5
	16	— 9·6		16	— 5·9
February	1	—13·5	August	1	— 6·3
	16	—14·2		16	— 4·4
March	1	—12·6	September	1	— 0·2
	16	— 9·0		16	+ 4·8
April	1	— 4·2	October	1	+10·0
	16	0·0		16	+14·2
May	1	+ 2·8	November	1	+16·3
	16	+ 3·7		16	+15·3
June	1	+ 2·4	December	1	+11·3
	16	— 0·4		16	+ 4·8

Time Definitions

Mean Time (MT): Clock time based on the average, or mean, solar day.

Apparent Time (AT): True time expressed by the position of the Sun.

Equation of Time (E): Difference between the values of Mean Time and Apparent Time. It constantly varies and the difference may amount to 16 minutes.

Standard Time (ST): The Mean Time in any of the world's standard time zones (Fig. 34). Each zone is represented by meridians 15° apart, one for each hour. The zones have their origin at the 0° longitude meridian at Greenwich.

Greenwich Civil Time (GCT): Local Mean Time (LMT) at 0° longitude.

Universal Time (UT): Greenwich Civil Time. Important as this is the time used in most astronomical observations and in navigation.

Julian Period (JP): A period invented to simplify the calculation of exact time between historical events. The period begins on 1 January 4713 B.C., and is counted as whole days independent of different calendar systems.

Julian Day (JD): The day number since the beginning of JP. The Julian Day begins at noon UT and continues through the night and is measured in 24 consecutive hours until noon of the following day. Especially useful in constructing light curves for variable stars.

Astronomical Day: A Julian Day beginning at noon UT.

Sidereal Time (ST): Literally 'Star Time'. The interval of time measured by use of stars as reference points. One Sidereal Day is the time taken by the Earth to rotate adjacent to a star =23 hrs 56 mins=4 minutes shorter than a solar day.

Variation in the Earth's Rotation Period

By accurate measurement of the Earth's rotation against a fixed point on the star sphere, it is possible to detect variations in the speed at which it turns on its axis. There are three kinds of variation known at present:

(1) A slow but steady increase in the length of the day.
(2) Seasonal variations.
(3) Irregular fluctuations – positive and negative.

The evidence for the first one dates back to lunar observations made by the English astronomer Edmund Halley in 1695, who decided that the effect was apparently a speeding-up in the Moon's motion. However, it is now known to be due to the Earth and results in a lengthening of the day by 0·0016 second per century. Although at first glance this seems a minute amount, if we use the Earth as a clock, it would amount to an 8-hour error in 4000 years.

Nowadays quartz and atomic clocks are used to keep accurate time, and these show that the Earth rotates more slowly in spring and more rapidly in the autumn. The maximum variation amounts to 0·0025 second (2·5 msec), but the amount can vary from year to year.

The irregular and seasonal variations are attributed to the redistribution of the air masses in the atmosphere, the movement of water in the oceans, and the movements within the Earth's mantle.

Length of Year

Because of the unpredictable irregularities in the Earth's rotation period, time is nowadays based on the motion of the Earth round the Sun. For the various year lengths in current usage see table below. For precise definitions see glossary.

YEAR LENGTHS

Year	Mean Solar Days	d	h	m	s
Tropical	365·242195	365	5	48	45
Sidereal	365·256360	365	6	9	10
Anomalistic	365·259643	365	6	13	53
Eclipse	346·620050	346	14	52	52
Gregorian	365·2425	365	5	49	12
Julian	365·25	365	6	—	—

Precession and Polar Wander

In *c.* 150 B.C., the Greek astronomer Hipparchus (190–125 B.C.) noted that the positions of the stars he had begun to recatalogue did not match up with the positions given in an earlier catalogue. At first he put down the differences to former inaccuracies, but he soon had to abandon this idea when he realized that the apparent displacements were uniform.

This was the discovery of the phenomenon that we call precession or – to give it its full title – *precession of the equinoxes*. This slow displacement of the star sphere, in relation to where the Sun crosses the celestial equator at each equinox, arises from the gravitational attraction of the Sun and the Moon tending to pull at the Earth's equatorial bulge. The action is opposed by the rotation of the Earth, resulting in an oscillation of the Earth's axis around the pole of the ecliptic which describes a circle 47° in diameter over a period of 25,800 years that has the effect of changing the pole star (*see* Fig. 6). This results in advancing the vernal equinox westward by 50″ per year, so that star coordinates which are measured from the spring equinox need to be referred to a particular year, or given epoch, e.g. 1920 or 1950, etc.

Superimposed on this motion is a short-term 'nodding' effect, or nutation, of the pole which occurs over an 18·6-year cycle. This is the result of the fact that the Moon's orbit is inclined 5° to the plane of the ecliptic and it tends to pull the Earth's equatorial bulge back into line. The effect was discovered in 1747 by the third British Astronomer Royal, J. Bradley (1693–1762).

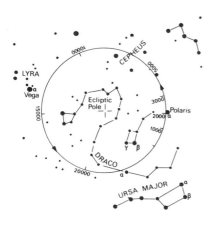

Fig. 6 The precessional shift of the north celestial pole round the pole of the ecliptic over 25,800 years.

There is yet another force present which causes a slow movement over a 14-month period, called the Chandler Wobble, named after an American who first measured it in 1891. This polar wander – or dance of the pole as it has been termed – is not fully understood, but it is thought to originate from the fact that the Earth is not a rigid body. A present-day theory attributes the wobble to a shifting in the axis of inertia of the Earth, which is the result of coupling forces between the fluid motion in the core and the surrounding rigid mantle – giving rise to linked electromagnetic forces.

The Earth as a Planet

The Earth is believed to have formed out of the primeval material 4600 million years ago. It is fairly typical of the smaller inner planets of the solar system, although it must be remembered that no planet can be an exact model of another owing to the different conditions prevailing at increasing distances from the Sun.

The density of the Earth is 5·52. It increases greatly towards its centre along with a considerable rise in temperature, at an approximate rate of 1 °F every 16 m (50 ft). The weight of the Earth has been calculated as $5·975 \times 10^{24}$ kg. The rocks forming the surface layer, or crust, total about 35 km (20 miles) in thickness, and consist of silicon, aluminium, iron, magnesium, calcium, sodium, potassium, hydrogen and other elements in smaller quantities. From our knowledge of the rest of the Universe the Earth can be considered fairly typical in its chemical make-up.

During the nineteenth century the source of the heat of the centre of the Earth was considered a mystery, but we now know that it is derived from the decay of certain radioactive elements that are locked away within the crust. Life in various primitive forms has existed on Earth for a considerable period of its history. Various dates have been derived in recent years for the oldest signs of life in the form of fossil remains. Structures found in the Biwabik iron formation in Minnesota are dated at 1700 million years old. In Queensland, Australia, fossil remains of algae are dated at 1600 million years. A claim has been made that hydrocarbon elements found in the Gunflint formations near Lake Superior, Canada, may be 2000 million years old. A still more tentative claim concerns microscopic structures found in pre-Cambrian rocks from South Africa, dated at more than 3000 million years. Yet another unconfirmed one is of a small squid-like fossil, about an inch long, embedded in the Pennsylvanian series rocks at Mazon Creek, Illinois, which at present holds the 'world record' with a quoted figure of 3100 million years.

Seismologists conclude that the Earth has a fluid iron core of about 3500 km (2200 miles) radius which is surrounded by a solid 2900 km

(1800 miles) thick silicate mantle on which the lighter 35 km (22 miles) thick continental crusts float like islands. It is now universally accepted that the continents have in the past drifted apart, and that the movement is still continuing in most places. About 200 million years ago South America and Africa were part of a single land mass known to geologists as Gondwanaland. The two continents fit exactly if matched by their continental-shelf structures which extend into the sea. Likewise Australia and Antarctica fit together almost exactly when their margins are taken at the 1,000-fathom level.

The majority of earth tremors, volcanoes and quakes are caused directly or indirectly by the internal forces of the Earth tearing the continental masses apart, the energy for which is derived from slow moving thermal currents upwelling within the crust. The action can be likened to conveyor belts of molten material driven up by convection from the mantle. Most of this material surfaces along the centres of the major oceans of the world and spreads across the sea floor before returning again to the mantle. The famous Mid-Atlantic ridge which runs southwards from Iceland through the Azores to the South Atlantic was thus formed by such action, like the similar ridge running down the central Pacific Ocean.

Shape of the Earth

Even by taking into account local height variations such as mountain chains and low-lying flood plains, the terrestrial sphere differs quite appreciably from the perfect spherical form so beloved by the early Greek astronomers.

Because the poles of the Earth remain stationary during rotation, a bulging is induced at the equator owing to the effects of the centrifugal force acting there. For the purposes of precise geodetic measurement and survey, the Earth is reckoned as a *geoid* – a name given to an idealized oblate body, sometimes called an oblate spheroid. The diameter measured round the poles equals 12,719 km (7900 miles), while the diameter round the equator measures 12,760 km (7926 miles).

Observational data derived from the deformation of artificial satellite orbits, in motion round the Earth, has given very detailed information about local variations in the Earth's shape. The orbits of artificial satellites are greatly influenced by bulges or depressions, and by noting the effects on satellite motion it is possible to interpret the information in terms of local height variations.

Results so far achieved have shown that the Earth is not gravitationally symmetrical about the equator. Its actual figure is slightly pear-shaped with the 'stem' directed at the North Pole. Total distortion is probably not more than 15 m (50 ft) and arises from the presence of two gravitational dimples; one is located near India, and the other off the west coast of North America.

The Atmosphere

The Earth is surrounded by a layer of gases which is termed the atmosphere from the Greek *atmos*, 'vapour' and Latin *sphaera*, 'sphere'. This layer has a profound effect on life over the Earth's entire surface and also on the way we see different astronomical phenomena.

All the Earth's weather patterns are triggered by the interaction of the incoming solar energy with atmospheric gases which sets up a complicated thermodynamic cycle. Without the influence of the Sun, there would be no climate. The fundamental pattern of the prevailing winds is on the whole brought about by the rotation of the Earth and the temperature differential that the Sun's radiation creates between the equatorial and polar regions. Local weather variations are more generally influenced by the distribution of sea and land masses.

The part of the atmosphere closest to the Earth is called the troposphere and extends to a height of about 13 km (8 miles). Next comes the stratosphere extending to 55 km (34 miles), followed by the mesosphere to a height of 85 km (53 miles). As we increase in height above the Earth's surface, the temperature drops. At the limit of the stratosphere it is about -55 °C. At a height of 25 km (16 miles), the temperature begins to rise again and reaches a value of -8 °C. Higher still, it begins to fall again, and at 94 km (56 miles) reaches a value of -80 °C. Beyond this point it rises again to give daytime temperature of 1220 °C, and about 750 °C at night. These high temperatures are known as kinetic temperatures and indicate the speed at which the molecules of the rarefied atmosphere are moving about.

With increasing height the density of the atmosphere falls off quite rapidly. At a distance of 5000 km there are still traces of atmosphere, but it is extremely tenuous and barely detectable. The density on the surface of the Earth is equal to 1.2×10^{-3} gm/cm^3. At 200 km (125 miles) the value has dropped to 10^{-13} gm/cm^3. The total pressure on a given square inch of the Earth's surface is about 14.7 lb/in^2 (= 1 atmosphere or 760 mm of mercury).

The gases comprising the atmosphere are a mixture of various chemical compounds. Their approximate composition, measured by volume is: diatomic nitrogen (N_2) 78 per cent; oxygen 21 per cent; argon 0.94 per cent; carbon dioxide 0.03 per cent; hydrogen 0.01 per cent; neon 0.0012 per cent; helium 0.0004 per cent; krypton 0.0001 per cent; water vapour 0.2–0.4 per cent and traces of other gases.

The atmosphere is a vital agent in maintaining a suitable temperature. It protects the Earth from high temperatures when the Sun is overhead and from low temperatures during the hours of darkness. It plays an important role in protecting the surface of the Earth from meteor-particle bombardment, for thousands of these tiny particles burn out in the upper atmospheric regions every day.

Carbon dioxide was probably much more abundant during the volcanic era of the Earth's history. At present more oxygen and carbon are combined in the rocks forming the Earth's crust than in the atmosphere. The ecological pattern of the living organisms of the Earth is very much dependent on a balanced equilibrium of atmospheric gases, and industrial pollution, which is increasing, may have a profound detrimental influence in the future.

From the astronomical aspect the atmosphere holds great significance in observational astronomy. The atmosphere absorbs or reflects a great deal of radiation coming from outside, with the consequence that it reduces the brightness of stellar objects and totally absorbs certain wavelengths of electromagnetic radiation. When we look at the sky, we are literally looking through an optical window with a limited view of the objects outside. At least 25 per cent of solar radiation is either scattered back or absorbed. The atmosphere does not absorb all the incoming light equally. It tends to allow red colours through more readily than blue, so that astronomical bodies appear 'redder' than they truly are. Most people are familiar with the blood-red appearance of the Sun at sunrise or sunset. At these times the solar light has to pass through a greater thickness of atmosphere to reach the observer, and more of the bluer light is scattered than when the Sun is overhead.

Another significant property of the atmosphere is its ability to refract or bend the light beam of an object coming to us from outside. This produces the effect of shifting its apparent position as we see it in the sky. For example, a star is made to look higher up than it actually is. When precise observations are required, correction tables have to be used to allow for this effect.

ATMOSPHERIC REFRACTION

Altitude	Refraction		Altitude	Refraction	
0°	34′	54″	13°	4′	5″
1	24	25	14	3	47
2	18	9	15	3	32
3	14	15	20	2	37
4	11	39	25	2	3
5	9	47	30	1	40
6	8	23	35	1	22
7	7	20	40	1	9
8	6	30	45	0	58
9	5	49	50	0	48
10	5	16	65	0	27
11	4	49	80	0	10
12	4	25	90	0	0

Earth Magnetism and the Aurora

The existence of a magnetic force within the Earth was known about for thousands of years before its true nature was discovered. The ancient Chinese and Mongols recognized that certain iron stones (lode stones) tended to point to a fixed direction on the Earth's surface, and it is hardly surprising that the Pole Star, which remains practically in the same position from night to night, was nicknamed the 'Lode Star'.

Every magnet has two poles. The Earth's magnetic poles are located very approximately in the direction of the geographical poles. However, they do not remain constantly at fixed points and describe small circles over a period of about 500 years. Over a much longer period, measured over several hundred million years, the magnetic poles also wander haphazardly across the entire Earth's surface, so that at times they may be located at 90° away from the axis of rotation. The present-day location of the north magnetic pole is lat. 76°N, long. 100°W, and the south magnetic pole lat. 66°S, long. 139°E.

Until recent times the cause and mechanism of the Earth's magnetic field was a puzzle. At present the generally accepted theory is that it arises owing to convection currents in the fluid core which act like a hydromagnetic self-exciting dynamo and produce an electromagnetic force. Geological evidence indicates that during the course of its 4600-million-year history, the Earth has switched polarity several times; so that the south pole became the north, and the north the south. These reversals can be studied through the modern science of palaeomagnetism which concerns itself with the measurement of weak fossil magnetism still remaining in certain rocks. Some rocks are known which provide evidence of being magnetized at a period when the Earth's magnetic field was in the very act of switching over its polarity.

There are two possible explanations of the causes of reversal polarity. It could have occurred in the rocks when they were cooling, as in the case of lava, by an unexplained mechanism which has the effect of suddenly switching the whole of the Earth's magnetic polarity. The alternative explanation provides for physical or chemical characteristics within a particular group of rocks which induce a spontaneous 'self-reversal'. Material with self-reversing characteristics is quite well known and has been synthesized in the laboratory, and in nature a few minerals do possess this property.

The Earth's magnetic field gives rise to lines of force exactly analogous to the force lines produced by iron filings and a bar magnet. These lines of magnetic force extend out from the Earth in curved paths and reach into nearer space, creating a powerful magnetic field well beyond the previously accepted limits of the Earth's atmosphere.

About 1000 km (600 miles) above the surface are encountered the Van

Allen belts, which are really extended parts of the true atmosphere and consist of energized ionized nuclei of electrons and atoms trapped by the Earth's magnetic field. These belts are named after the American physicist, James A. Van Allen, who detected them from information relayed back by the early satellites. It is known that the Van Allen belts (Fig. 37) may be dangerous to living organisms not protected by adequate shielding such as lead sheeting. Their origin is attributed to the bombardment of Earth by particles shot out from the Sun via the massive explosive solar flares. However, cosmic-ray collisions may also be a contributing factor, and nuclear explosions carried out at great height might have their long-term effect if such tests were routinely carried out. The nuclear explosion of November 1958, which occurred at a height of 160 km (100 miles) above the Pacific, caused a temporary build-up of super-intense radiation belts.

When a particle from the Sun reaches the Earth, it becomes captured by the Earth's magnetic field. The particle is forced into a spiral orbit and, continuing a rotary action, it may shuttle from one hemisphere into the other before finally decaying or leaking away. The lifetime of particles is usually measured in hours, but they can survive many days if caught within the central part of the radioactive belt (Fig. 37).

The charged particles reaching the Earth from the Sun also give rise to the brilliant and often spectacular auroras. These are caused by the particles being funnelled into the atmosphere in the polar regions along the force lines of the Earth's magnetic field. The auroras are particularly active during the period of sunspot maximum when frequent giant flares occur that can be seen to curve out thousands of miles away from the solar surface (Fig. 29).

As the charged particles (chiefly electrons) penetrate the upper atmosphere, they ionize its tenuous gases which finally become so excited that they give rise to brilliant sky displays or glows. The colours observed are generally green, blue, white or red. Spectrum analysis reveals the presence of bright emission lines which originate from oxygen, nitrogen and sodium atoms/molecules of the upper air.

At sunspot maximum the displays occur over wide zones round each pole at heights ranging from 80 to 300 km (50 to 200 miles). More rarely they also occur well beyond this distance. Displays are usually observed more frequently above 50° (solar latitude) in high temperate latitudes in both hemispheres. But at times of great intensity they can also be seen in the equatorial region.

When auroral displays occur in the northern hemisphere, they are called Northern Lights (Aurora Borealis), and in the southern hemisphere Southern Lights (Aurora Australis). The forms most frequently seen consist of ray-like structures, ill-defined glows and areas of curtain-like features which are all very impressive to watch as they rapidly change their form and character (Fig. 38). With suitable fast colour films and modern 35 mm single-lens reflex cameras they can be photographed quite easily. In remote geographical locations, away from extraneous noise, sounds of a peculiar kind have often been reported accompanying brilliant auroral displays. Although they are not yet confirmed positively, they could well be a manifestation of the little-understood electrophonic noises similar to those often heard during the fall of a meteorite or the flight of a brilliant fireball. These noises travel ahead of the meteorite and have been described as 'like the playing of the wind in telephone wires' or 'swishing' sounds, and animals appear to be particularly receptive to their audible frequency range.

The Calendar

The word calendar is derived from Roman times and comes from the Greek *kalend* which has the literal meaning 'I cry', a leftover from an age when the town crier announced the orders of the Pontifices – such as when market days were to be held and the beginning of each month.

The Roman month began when a new Moon appeared as a thin crescent immediately following the setting Sun. Hence we have 'moon time', *moonth*, from which we obtain the present-day month. However, this Roman calendar was not very accurate. A great deal of juggling had to be done since the months were divided into alternate 29- and 30-day intervals. Over a period of time these intervals became out of phase with the cycle of the Moon's phases.

All the early calendars were based on some kind of natural cycle or rhythm. The first Egyptian calendar, based on the female cycle, has survived in modified form to the present day. It was supposedly devised by Thoth about 5000 years ago, and the legend is recorded in the writings of Plutarch as it was told to him by the chief priests of an ancient Egyptian temple.* The gist of the story is that Thoth became interested in devising a calendar which would fit the life-cycle of both animals and people and take into account the *true* length of the year.

It has often been stated that the subsequent Egyptian calendar is the only intelligent calendar which has ever existed in human history. A year consisted of 12 months of 30 days each, and 5 additional days at the end of each year. A more scientific explanation of its origin is that the Egyptian year began with the heliacal rising of the bright star Sirius (α CMa). The contemporary Egyptians called this star Soth, and the calendar based on it is referred to as the Sothic year or the Sothic cycle. The heliacal rising of a star

* Sir James George Frazer, *The Golden Bough* (New York, 1936), vol. II, p. 6.

is defined as the rising of a star just observable in the morning twilight before the Sun rises. It has been inferred that the Egyptians could observe Sirius when the Sun was $10\frac{1}{2}°$ below the horizon.

The day when the Sothic calendar was adopted has been reckoned as approximately 4250 B.C. But the 365-day calendar is really too short, since the seasons shift through the months in about 1460 years in an ever-repeating cycle. To overcome this difficulty an extra intercalary day was suggested in 238 B.C. to bring the year into coincidence with the seasons. This extra day was the basis of the leap-year rule introduced later by Julius Caesar.

The Egyptian year was divided into three seasons of four months each. Although it was astronomically based, it was fundamentally a calendar constructed for practical agricultural purposes. Sirius was merely a starting point because of its convenient rising just before the annual Nile flood.

The Babylonians depended on a strict lunar calendar which was very difficult to operate. This was also used by the later Greeks who occasionally introduced into it political considerations which influenced its day-to-day chronology and resulted in shortening or lengthening a particular month. As a consequence, historians often have great difficulties tracing back events in Babylonian and Greek times, whereas for Egyptian times the problem is a fairly straightforward one: simply multiply the number 365 by the number of years required. It is not surprising then that by the year 45 B.C., Julius Caesar, after advice from the Alexandrian astronomer Sosigenes, brought about a long-needed reform. He separated the length of the month completely from the phases of the Moon and gave to each a fixed number of days. His year contained 365 days, with an extra day (an idea borrowed from the Egyptians) included every fourth year. This calendar has now been in use for nineteen centuries, but in 1582, Pope Gregory XIII introduced a new rule for fixing leap years, to bring the calendar into closer agreement with the true passage of time. This change was not introduced into England and the American colonies until 1752, but it was adopted by Scotland in 1600.

Pope Gregory's change came about from the meeting of the Church Council at Nicaea in A.D. 325 when the problem of the date for Easter observance was first discussed. At the Church Council held in Italy 1220 years *later*, the subject was again on the agenda, since between these years the existing calendar had developed an error totalling 10 days. When Gregory's reforms came into effect in October 1582, the date 4 October was followed by 15 October. When the change became finally adopted in England and America (1752), 2 September was put down as 14 September. An adjustment of 11 days was needed owing to the year 1700 being a leap year by the Julian system. Dates prior to the change are referred to as Old

Style (O.S.) and dates since, New Style (N.S.). Gregory's change has caused many a headache to European historians, for most Protestant countries other than England and America did not adopt this until many years later. In Protestant Germany it was adopted in 1699, Ireland 1788, Turkey 1917, Rumania 1919 and in Greece not until 1923.

The Mohammedan calendar is still based on a pure lunar year. The day begins at sunset. Each week has seven days which continue independently of the months. The Mohammedan years are numbered from the time of the flight of Mohammed from Mecca in A.D. 622, 16 July.

The Jewish calendar is based on a combined luni-solar year, but from A.D. 338 the beginning of the month was fixed by calculation rather than the first sighting of the crescent Moon. New Year begins at 6 p.m., and the years are numbered from the time of the 'world creation' – assumed to be 3761 B.C., 7 October.

In the Western calendar the use of the terms B.C. and A.D. were introduced about A.D. 607. The important date of Easter was fixed by the Church Council of Nicaea so that it could be celebrated on the same date throughout the Christian world. It is fixed to occur on the first Sunday *after* the full Moon *after* the spring equinox, which means that its calendar date is never the same two years in succession.

The *Book of Hours* (Fig. 41), designed as a medieval calendar prayer book, is both a scientific document and one of the first examples of modern realism in painting. It was made for the Duc de Berry (*c.* 1400) and painted by Pol de Limbourg and his brother to sustain the interest of the Duke while at prayer in church. Each month shows a picture of the various châteaux belonging to the Duke, and it also contains all the saint days and festivals significant during the period. The semicircular field at the top is the most important part from the astronomical point of view for it reflects the basic facts of the Sun's anomaly, i.e. it travels more slowly in its orbit in summer and faster in winter. Unfortunately, for both art and science, the twelve folios were never completed, as the brothers Limbourg left the court when the Duc de Berry died in 1416.

In order to appreciate the difficulties which beset the ancient calendar inventors, consider the odd-number relationships in three important quantitative elements contained in the two fundamental astronomical cycles:

The solar year	= 365·2422 days
Lunar month	= 29·53059 days
Lunar months into solar year	= 12·36827

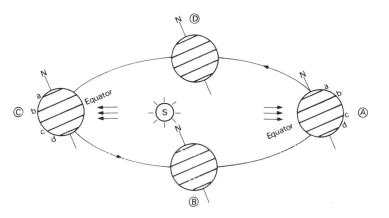

Fig. 7 The seasons:

(a) The position of the Earth during the four seasons. Owing to the eccentric nature of the Earth's orbit round the Sun, summers and winters in the southern hemisphere are more accentuated than in the northern.

A Summer, northern hemisphere; winter, southern hemisphere.
B Spring, northern hemisphere; autumn, southern hemisphere.
C Winter, northern hemisphere; summer, southern hemisphere.
D Autumn, northern hemisphere; spring, southern hemisphere.

Key: a Arctic Circle; b Tropic of Cancer; c Tropic of Capricorn; d Antarctic Circle.

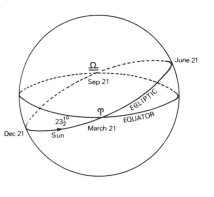

(b) The seasonal path of the Sun in relation to the celestial equator. The vernal equinox (♈) occurs about 21 March, and the autumnal equinox (♎) about 21 September.

The Seasons

Because of the Earth's motion round the Sun during the course of the year and the tilt of the polar axis amounting to $23\frac{1}{2}°$, seasonal variations in climate occur over a large portion of its surface area. The weather pattern of the Earth is principally determined by the total amount of solar radiation that enters the atmosphere, but it is also locally influenced and complicated by the distribution of land and sea.

Seasonal temperature variations are brought about by the fact that radiation falling on a particular area varies according to the angle presented to the Sun. It will be noted (Fig. 7a) that the angle made by the Sun's rays on the Earth in any geographical spot depends on the Earth's location in its orbit. It must be remembered that the Earth's polar tilt, owing to gyroscopic action, remains pointed in the same direction in space throughout the year.

The apparent shifting of the Sun, as seen from the surface of the Earth, north or south across its surface, is defined at the northern limit by the tropic of Cancer, and at its southern limit by the tropic of Capricorn. It follows that since the Earth's tilt is $23\frac{1}{2}°$, each tropic lies exactly at $23\frac{1}{2}°$ north or south of the equator. The Arctic and Antarctic circles define the north and south latitude parallels where the Sun does not rise above the horizon for certain periods during each hemisphere's respective winter season.

The seasonal periods then are divided according to the apparent motion of the Sun in relation to the Earth's surface. The equinoxes – meaning *equal day and equal night* – occur when the Sun appears to cross the celestial equator from south to north or north to south. At the spring (vernal) equinox (northern hemisphere), on 20, 21, or 22 March, the Sun appears to pass from south to north across the celestial equator. At the autumnal equinox, 20, 21, or 22 September (northern hemisphere), the Sun appears to pass from north to south. The date when the Sun is furthest north overhead at the tropic of Cancer at midday, is termed the summer solstice (near 21 June, northern hemisphere) – meaning *Sun standing still*. And the date when the Sun is overhead at the tropic of Capricorn, is termed the winter solstice (near 21 December, northern hemisphere). In the southern hemisphere it must be borne in mind that winter and summer are reversed to the seasons of the northern hemisphere.

The Celestial Sphere

The first watchers of the sky imagined the celestial sphere as a roof, or vault, on which the heavenly bodies were studded and represented as lying together side by side on its inner surface. Even Omar Khayyam, the astronomer-poet of Persia (1017–1123), spoke of 'that inverted bowl we

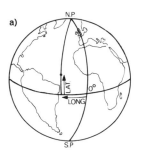

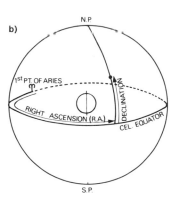

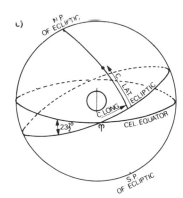

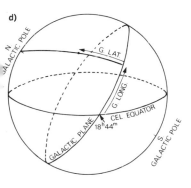

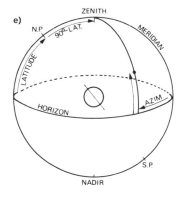

Fig. 8 The coordinate systems:

(a) The Earth system.
(b) The equatorial system.
(c) The ecliptic system.
(d) The galactic system.
(e) The horizon(tal) system.

33

call the sky'. Today, although our ideas about the Universe are different, we utilize this same convenient and simple concept when we use the coordinates that fix the position of celestial objects in relation to the Earth in space (Fig. 8).

Positions of celestial objects, stars, comets, planets, etc., are fixed in exactly the same way that we define a position on the Earth's surface. Any location on Earth can be specified by referring to its latitude and longitude. Latitude is measured north or south of the equator ($0°$=equator, $90°$ the poles). Longitude is measured east or west from an imaginary line, or great circle, passing from the North Pole, due south, through the position of the old Greenwich Observatory in London and on towards the South Pole. This is often referred to as the *Greenwich Meridian*. Longitude can be expressed in angular measure: $°$, $'$, $''$, or in time: hrs, mins, secs. The diameter of the Earth $= 360° = 24$ hours.* Longitude actually expresses the time of any geographical location in relation to the time of the Greenwich Meridian, or great circle, which acts as basis for time anywhere in the world (Fig. 8a).

The celestial equivalents to longitude and latitude are called *right ascension* (RA) and *declination* (Dec). Similar to longitude, right ascension can be measured in either time (hrs, mins, secs) or angular measure ($°$, $'$, $''$). Declination, like latitude, is measured in degrees north or south of the *celestial equator* (equator$=0°$; north celestial pole $+90°$; south celestial pole $-90°$). Although one occasionally comes across N and S declinations, the official prefixes are $+$ (plus) for north and $-$ (minus) for south.

Whereas longitude on the Earth is measured from a zero point, i.e. the great circle of Greenwich, right ascension in the sky is measured from the *First Point of Aries* (Fig. 8b). This is the position of the Sun as it crosses over the celestial equator at the time of the spring (vernal) equinox (Fig. 7b).

Astronomy employs a number of different coordinate systems on the celestial sphere. One has to remember which system is being specified for any particular system of coordinates. The one most commonly met with is called the *equatorial system* (Fig. 8b), which is almost identical to the longitude and latitude system on the Earth's surface. The second most frequently encountered is called the *ecliptic system* (Fig. 8c). This is similar to the equatorial, but instead of measurements being related to the celestial equator they are related to the ecliptic – or the apparent path followed by the Sun, Moon and the planets as they move across the sky. The ecliptic is inclined to the celestial equator at $23\frac{1}{2}°$, owing to the Earth's axial tilt. To avoid possible confusion the ecliptic system uses the terms *celestial longitude* and *latitude*, both quantities being measured in angular $°$, $'$, $''$.

* e.g. 1 hour of time $= 15°$ angular measure.

Both systems have their particular uses and advantages. No one knows for certain who invented them. Hipparchus, the Greek astronomer, appeared to know about both methods, so they must have originated before 200 B.C. One of the earliest artefacts known for measuring star angles was found in the tomb of Tutankhamen in 1923 and dates from the fourteenth century B.C. Generally speaking the equatorial system is the one most frequently used. All modern star charts, maps and catalogues use it to define the location of celestial objects.* The ecliptic system is used for constructing planetary tables, since by using the ecliptic, the approximate position of the planets can be denoted by using one coordinate only, i.e. celestial longitude.

The *galactic system* (Fig. 8d) is occasionally used in connection with problems relating to the structure of the Milky Way (or Galaxy). Another important coordinate system, often also used by surveyors, engineers and navigators, is the simplest one of all, called the *horizon(tal) system* (Fig. 8e). In this the two coordinates are: *azimuth* measured round the horizon from the observer's true south point – westwards;† and *altitude*, the angular measurement above the observer's horizon (horizontal plane). This system also includes the terms *zenith*=point directly overhead, and *nadir*=point directly below. *The meridian* is a north–south line (great circle) passing through the observer's zenith.

The equatorial system utilizes sidereal time (star time). This is the star time indicated by the observer's meridian (N–S line). Its zero point begins at the First Point of Aries=RA 0^h 0^m 0^s. The sidereal time then is the difference between the First Point of Aries and the observer's meridian. When the First Point of Aries is exactly on the observer's meridian, the sidereal time is zero. Because right ascension, like sidereal time, is reckoned from the First Point of Aries, the right ascension will also be zero. The *hour angle* of a star is the sidereal time which has elapsed since the passage of a star across the observer's meridian. It will be seen that sidereal time *minus* right ascension=hour angle.

Finally, one may occasionally encounter *heliocentric coordinates*. These are coordinates used as relating to the Sun's centre. They are used in some kinds of mathematical astronomy such as orbit computing, and they can easily be converted to *geocentric coordinates* (Earth centre coordinates) by simple trigonometric formulae, which are to be found in most astronomical mathematical text-books.

---

* For example, the bright star Sirius (α CMa) RA=6^h 43^m Dec$-16°$ $39'$. Turn to p. 184 and locate it, using these coordinates.
† Navigators, engineers and surveyors measure azimuth from true *north*.

Radiation

From a location on the surface of the Earth all knowledge of the Universe comes to us in the form of electromagnetic radiation. If we are using the eye alone, what we see is only a part of the electromagnetic spectrum called visible light. The nature of light is not fully understood. One of its paradoxes is the dual property of acting both as a wave pattern and in a corpuscular (particle) form. Nevertheless, we know a good deal about its other physical qualities; and through measurement, recording techniques and conjecture we can formulate a coherent picture about the various astronomical bodies outside.

In its broadest concept electromagnetic radiation is closely associated with the atomic structure of matter. The structure of an atom is *visualized* in the form of a central nucleus round which revolve electrons. Waves, or particles, of energy are created when an atom gives up, or releases, one or more of its orbital electrons, or if the nucleus itself is radically altered. Electromagnetic radiation is transmitted through the vacuum of space; it can be refracted (or bent) or reflected, and its direction is affected by strong gravitational fields.

Even the ancients knew that light must travel at very great speed. Galileo tried to measure its velocity by exchanging light signals using lanterns between two widely spaced points, but his crude attempts met with complete failure. It was not until the Danish astronomer Roemer, in the seventeenth century, that the first accurate estimate was known. Because light travels at tremendous velocity, it cannot be measured by ordinary methods. It was only via the information gleaned by observing the irregularities in the motion of Jupiter's satellites that Roemer was able to arrive at a figure very close to the presently accepted figure of 300,000 km (186,000 miles) per second. When the Earth is on the same side of the Sun as Jupiter, light takes less time to reach the Earth than when we are on the opposite side to Jupiter. Since the time difference between calculated and observed positions is known, and the differences in distance are known, the velocity computation is a simple one. Nowadays there are many highly sophisticated methods available to arrive at the velocity.

Visible light represents only a very narrow band in the known range of electromagnetic radiation (Fig. 42). The visible range is diminished even further by the filtering effect of the atmosphere. Human eyes are sensitive to a range of wavelengths, 4000 Å to 6800 Å.* Outside the ozone layer in the atmosphere the orbiting observatories can record visible light below 2900 Å. Some animals are able to see larger or smaller wavelengths than are normally visible to human eyes. Bees are thought to see ultra-violet

* Å = Ångstrom unit, used for measuring the wavelengths of light. 1 Å = 10^{-7} mm.

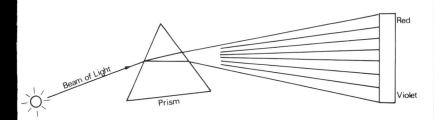

Fig. 9 Refraction of light through a prism to produce a continuous colour spectrum. A similar effect may be achieved by substituting a grating inscribed with a network of fine lines.

light at 3000 Å, but their vision is cut off in the longer red wavelengths.

When a beam of light is passed through a prism, or refracted or reflected off a grating inscribed with a network of fine lines, the beam is split up into the familiar rainbow colours: red, orange, yellow, green, blue, indigo, violet. This spectrum sequence, extending from long wavelengths to short ones, can be remembered by the mnemonic: Richard Of York Gave Battles In Vain (Fig. 9)*. It was Isaac Newton, in 1668, who first discovered the properties of glass to *refract* white light. Later the Englishman Wollaston and the German Fraunhofer noted that when white light was first passed through a narrow slit, darker lines were rendered visible in the various colours of the spectrum (Fig. 43). Later the German physicists G. R. Kirchhoff (1824–87) and R. W. Bunsen (1811–99) – of Bunsen burner fame – were able to identify these dark lines as belonging to various chemical elements contained within the light source. Nowadays 92 naturally occurring elements are known plus others artificially made. From these early beginnings has arisen the modern science of spectroscopy, which, using either prisms or gratings to split up the light, in combination with various laboratory techniques, can tell us a great deal about the physical and dynamical nature of stars that we can only observe visually as point sources. Thus astronomical spectroscopy (*see* p. 239) is the foundation stone of astrophysics. However, we are now able to utilize other non-visible parts of the electromagnetic spectrum. The energy waves utilized by radio telescopes lie in the invisible part of the spectrum – as do X-rays, ultraviolet and infrared emissions. We can detect stars and other similar cosmic objects which have no visible emissions. The X-ray stars most certainly exist as discrete bodies even though our eyes deceive us. It is no longer reasonable to adopt one of the ancients' philosophical arguments: *If I cannot see: I do not believe.*

* An American version: Read Out Your Good Book in Verse.

II THE SOLAR SYSTEM

Nature of the Solar System

All the celestial bodies such as planets, satellites, comets, meteors, fireballs and meteorites which we observe in the vicinity of the Sun are members of the solar system. All these bodies are acted upon by the gravitational field of the Sun and describe various shaped paths or orbits round it. Some of the bodies, like the satellites of the planets, describe orbits round their parent planets, but they are still, at the same time, orbiting with the planets round the Sun.

From the earliest times skywatchers had noted that certain 'stars' were not fixed, but that they moved or wandered across the heavens in relation to the others. Thus the word planet is derived from the Greek *planetes*, 'to wander'. Although several Greek, and probably some earlier Babylonian astronomers, placed the Sun at the centre of the system of planets, Ptolemy's geocentric ideas (*see* p. 13) remained the accepted view until Copernicus rediscovered the heliocentric system in the sixteenth century.

Galileo Galilei (1564–1642) formulated some simple dynamical laws about falling bodies which described the basic laws of gravity on the surface of the Earth. Later both Kepler and Newton formulated complementary sets of laws which explained the motion of the planets (*see* p. 16).

The ancients knew of five planets: Mercury, Venus, Mars, Jupiter and Saturn. The others were discovered after the invention of the telescope; Uranus in 1781, Neptune in 1846 and Pluto in 1930. Apart from the larger planets there exist a group of minor planets, or asteroids, whose orbits lie between Mars and Jupiter. The first of these smaller bodies was discovered in 1801, and nowadays we know that their total population probably exceeds 30,000.

The fundamental difference between a planet of the solar system and a star is one of physical characteristics. The Sun, our nearest star – which we are all sometimes inclined to forget – is a hot gaseous body which obtains its energy from thermonuclear reactions taking place in its interior. In many ways it resembles a continuously acting hydrogen bomb, for the large releases of energy, from which it derives its light and heat, are produced in a similar manner. The Sun is self-luminous, and we are able to see it as a consequence of its own radiation. The planets, however, are cool, non-luminous bodies that generate insufficient energy within their interiors to render them visible in ordinary wavelengths of light. We can only see the planets by the light reflected back off their surfaces, or atmospheres, which is first radiated by the Sun. These definitions certainly hold good for planetary members belonging to the solar system, but elsewhere in space, dense, non-luminous bodies may in fact be bizarre varieties of stars which do not generate sufficient energy to render themselves visible, or they are going through astrophysical processes not yet understood.

Planetary Movements

From our position on Earth the planets appear to move across the sky confined to a particular zone called the Zodiac. This is because all the planets revolve round the Sun in orbits approximately in the same plane (like a flattened dish) with the exception of some of the minor planets whose orbits are steeply inclined. Since the Earth's axis is tilted at $23\frac{1}{2}°$, this plane or ecliptic zone, also appears to be tilted in relation to the celestial equator, and they only coincide at the time of the equinoxes (see p. 32).

The orbital motion of all the planets is termed *direct*, or contrary to the movement of a clock (anti-clockwise). The planets Mercury and Venus, which have orbits inside the limits of the Earth's orbit, are called *inferior* planets; while all the planets which have orbits outside the Earth's orbit are called *superior* planets.

All the planets have different orbital velocities which decrease outwards from the Sun, so that Mercury travels fastest and Pluto is the slowest. Because we observe all the other planets from our own moving Earth, they sometimes appear to describe odd movements such as 'backward', retrograde loops, or if plotted against the background of stars, they may also appear to remain stationary for some days. Figs. 10 and 11 show how this apparently contradictory motion comes about for both an inferior and superior planet.

Planetary Configurations

As a planet revolves in its orbit, it will appear at different locations in the sky in relation to the Earth and Sun. These positions can be described with a precise nomenclature both for inferior and superior planets (refer also to Fig. 11).

Opposition: Superior planet in line (or nearly in line) with Earth and Sun with the Earth between.
Quadrature: Superior planet at right angles to Earth and Sun.
Conjunction: Superior planet in line with Earth and Sun, but beyond the Sun.
Greatest elongation: Inferior planet at right angles with Earth and Sun.
Inferior conjunction: Inferior planet in line with and between Earth and Sun.
Superior conjunction: Inferior planet in line with Earth and Sun, and beyond the Sun.

a)

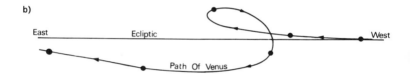

Fig. 10

(a) Planetary motion: inferior planet
(b) Retrograde loop of Venus.
(c) Planetary motion: superior planet (Mars).

b)

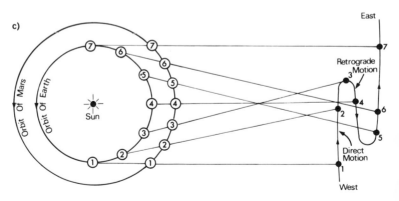

c)

40

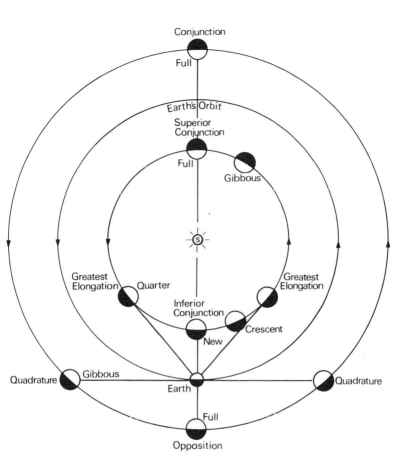

Fig. 11 Inferior and superior planet configurations.

Planets may also be in conjunction with one another or with the Moon. The pseudo-science of *astrology* attaches great importance to conjunctions, but in *astronomy* they have quite a different significance.

When a planet is in *conjunction*, it is on the meridian (the observer's north–south line) at noon. When it is at *opposition*, it crosses the meridian at midnight. At the time of opposition a superior planet is very suitably placed for telescopic observation – particularly so in the case of Mars. When a superior planet is in quadrature, its meridian passage is either 6 a.m. or 6 p.m.

Inferior Planets

Because the orbits of Mercury and Venus are within the orbit of the Earth, they never appear to wander far away from the vicinity of the Sun. Both planets are therefore best observed as early morning or early evening objects. Mercury will never be further than 28° away from the Sun and Venus 48°.

It is not surprising that they were called morning and evening 'stars' by the ancients, but it was some time before they realized that they were the same bodies only on different sides of the Sun. Also because they lie within the orbit of the Earth, both planets show distinctive phases exactly like the Moon's phases (Fig. 12). At the time when they are closest to the Earth, they may also transit, or cross the disc of the Sun. However, because of other orbital factors, these are comparatively rare events. The last transit of Mercury took place in 1970, and the next is due to occur on 13 November 1986. The transits of Venus, which occur in pairs, are even rarer occurrences. The last time was in 1874 and 1882, and the next will not be seen until 2004 and 2012.

Superior Planets

Since all the superior planets lie outside the Earth's orbit, they can be seen in any quarter of the heavens along the ecliptic plane. Although we see most of their illuminated surfaces, in certain orbital configurations they have a slightly gibbous phase, particularly in the case of Mars.

Periods of the Planets

The period of revolution of a planet round the Sun can be expressed either as:

(*a*) *Sidereal period:* The time a planet takes to make one complete revolution round the Sun from a particular point as seen from the Sun and back again.
(*b*) *Synodic period:* The time taken by a planet to go from a given elongation back to the same elongation again.

Fig. 12 Phases of an inferior planet (*see also* Fig. 11).

The synodic period is the period we observe from the moving Earth and is used to determine the sidereal period by the simple relationships:

superior planet : $1/S = 1/E - 1/P$
inferior planet: $1/S = 1/P - 1/E$
when E = sidereal period of the Earth
when P = sidereal period of the planet
when S = synodic period of the planet

The Sun

The Sun, because of its colossal mass, is the most dominant member of the solar system. When we observe it through a telescope, or rather if we project the image on to a white card to avoid damage to the eye, what we see is not a solid rigid surface but rather the outer layer of a large sphere of gas. Because of the properties of the gas we cannot see much below the surface layer called the photosphere – or really the limit of the part which presents a sharp disc in ordinary photographs. The diameter of the Sun as defined by the photosphere is 1,392,000 km (865,000 miles).

The distance between the Earth and the Sun averages 150 million km (93 million miles), but it varies because the Earth revolves in an elliptical orbit, with the Sun at one focus, so that the Sun is nearer to us in the northern winter than in the summer. This distance, called one astronomical unit (1 A.U.), is one of the most important units in astronomy. It provides half the trigonometrical base line from which we are able to measure the distance of the nearer stars.

The volume of the Sun exceeds the Earth by 1,300,000 times. Its mass is about 750 times the combined mass of all the other bodies in the solar system. The average density is a little greater than water, which is about one quarter of the Earth's density; but lower in the Sun's interior, it rapidly increases so that the hot gas becomes nine times greater than the density of iron.

Sunspots

The surface of the photosphere usually shows one or more irregular dark blotches or spots accompanied by smaller ones scattered about them. They can often be seen by the naked eye, particularly when the Sun is low down near the horizon and shining through dense mist or fog. There are numerous references to them in ancient Chinese, Japanese and Korean annals, and they were also supposedly observed by the Greek Theophratus in 300 B.C. In more modern times they were rediscovered telescopically by Galileo, in July 1610. However, at first he was not convinced of their reality and held back his announcement until April 1611, about the time they were also independently discovered by Fabricius of Wittenberg and Scheiner, a Jesuit priest.

Galileo evolved a method of observing sunspots which did not incur damaging the eyes. The method is still in vogue today, and is used both in visual and photographic observations by projecting the Sun's telescopic image beyond the eyepiece of the telescope and focusing it on to a white card or plate (Fig. 44).

In common with its attendant planets, the Sun also rotates on its axis. The sunspots enable the rotation period of the Sun to be determined. At the equator the Sun rotates in about 25 days in an east to west direction, or the direction of motion taken by the planets as they *revolve* round the Sun. In 1860, Carrington discovered that the rotation period varies as might be expected with a 'fluid' gaseous body. Between the equator and the Sun's polar regions, the period gradually lengthens to 34 days. The solar rotation axis is slightly tilted at an angle of $7° 15'$, so that spots appear to move across the Sun's disc in varying directions at different times of the year.

Solar Energy

How does the Sun derive its great heat? This was indeed a puzzle to the physicists of the nineteenth century and the period of the early twentieth, before nuclear and thermonuclear reactions were understood. The age of the Sun is also important, since it helps to estimate the age of the Earth and supplies answers to many other important astronomical and general scientific questions. In the nineteenth century, Lord Kelvin considered the Sun could be no older than 20 million years, and he was confident that it

44

would quickly burn itself out. Thinking had not progressed beyond the idea that it must be made up of some kind of combustible matter analogous to coal. As a consequence, it was easy to calculate and estimate how long it would continue burning.

Charles Darwin's evolutionary theory required a much longer time-scale to manifest itself than Kelvin's ideas would allow. For a long uncertain period, the physicists were responsible for casting doubt on the brilliant evolutionary evidence put forward by the zoologists and geologists. Poor Darwin himself was greatly troubled by the problems surrounding the age of the Sun. In 1868, he wrote to his great friend Lyell, a geologist: 'I take the Sun very much to heart. . . .'

Fortunately, however, atomic physics was not too far away, and in the twentieth century, helped by the modern sciences of genetics and palaeontology, it eventually vindicated the longer time-span required by the Darwinian theory.

Modern atomic theories allow for a much greater age for the Sun, and, equally significant, a much longer future. The generation of energy in the Sun's interior is almost exclusively due to nuclear processes and it radiates as much energy every second as would be released by the simultaneous explosions of several billion atomic bombs. Or to quote another analogy: it equals the bituminous heat from 11 thousand million million tons of coal. The Sun is an extremely hot body with a surface temperature of about 6000 °C. Temperature increases rapidly towards the centre to reach 20,000,000 °C. Geological evidence indicates that the Sun's radiation has been fairly constant for more than 1000 million years and it would appear to be in a stable condition. The difficulty of appreciating the energy released by atomic processes is that our minds are still attuned to old-fashioned mechanical ideas. For example, when we are told the Sun is a spherical mass of incandescent gas, the mind conjures up a picture based on familiar everyday experience. Can we truly appreciate that even if the Sun were the same density throughout, the pressure at the centre would be a billion atmospheres, and yet it still remains a gas? The temperature is so intense that the atomic nuclei split. When matter attains such a state, it is called a plasma. In this unfamiliar environment of the Sun's hot interior, the atomic nuclei collide with great frequency at high speeds, and various reactions occur between them.

Most of the Sun is made up of hydrogen gas. The fragmentary atomic nuclei of hydrogen atoms forming the plasma interact by a series of reactions which finally result in the transformation of hydrogen into helium. This is named the proton-proton reaction. Briefly, the atom of hydrogen is of the simplest construction and consists of a single proton surrounded by a single 'planetary' electron. In the hot interior of the Sun the electron is stripped from the nucleus and the resulting protons coalesce through a series of successive collisions to form helium nuclei (α particles). As the process continues, the internal hydrogen is gradually used up, and the helium becomes abundant. During these processes energy is released in a fashion partly similar to that which occurs during ordinary combustion. The analogy, however, can only be applied loosely, for the nuclear processes are infinitely more energetic than ordinary burning processes. Thus all the electromagnetic radiation in the form of heat and light which the Sun pours out during its life-span of some thousands of millions of years is so produced. Nevertheless, it must obviously have a measurable life-span, and indeed, like other stars, the Sun slowly evolves from one state to another. This story of stellar evolution can be followed in Part III.

The Solar Cycle

In 1843, a German amateur astronomer, H. S. Schwabe, announced to the world that he had detected a definite periodicity in sunspot frequency. Schwabe, over a period of 43 years, had observed the Sun on every possible occasion with a small telescope. The combined results indicated quite conclusively that in a regular 11-year cycle the sunspots increased towards a maximum and then fell away to a minimum during the following years (Fig. 13).

Sunspots occur in well-defined zones between 8° and 35° north and south of the Sun's equator. The spots of a new cycle gradually move towards the equatorial zone as the cycle progresses, while spots from the previous cycle are still present a few degrees north or south of the equator.

Fig. 13 The sunspot cycle. The frequency of sunspots varies from year to year over an 11-year cycle. The graph shows the total number of visible spots each year in the period 1925 to 1965.

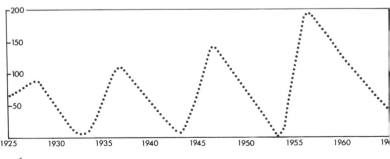

As a consequence we see two separate 11-year cycles overlapping.

Although the sunspots appear as completely dark areas (called *umbrae*) surrounded by less dark zones (called *penumbrae*), they are nevertheless extremely hot areas. They only appear dark by *contrast* in comparison with the much hotter surrounding area. Measurements indicate that the dark area of a spot is about 40 per cent that of the surrounding photosphere. The Sun possesses a very strong magnetic field which is particularly well developed round sunspots. Both a north and a south pole are present in each spot complex, and there are very precise laws governing the arrangements of polarity in each hemisphere. With the commencement of a new cycle of spots, the polarity becomes reversed so that in terms of magnetic activity the Sun has a cycle twice the 11-year period, or 22 years in length.

The nature of sunspots is not fully understood. The largest may be over 200,000 km (125,000 miles) across and plainly visible to the naked eye. The smallest are only a few thousand kilometres across and appear like small pores in the Sun's surface. Spots continually change their appearance and size, and some may persist for many revolutions of the Sun. Through a telescope they give the impression of vortex-like structures. With the spectrohelioscope, which shows the structures only in hydrogen or calcium light, one sees a whirling motion of gases in an up-and-down circulatory movement. Theory suggests that sunspots are due to magnetic hydrodynamic phenomena originating in the interior of the Sun. This is a kind of coupling action between magnetism and fluid forces (since the Sun's gas can be reckoned to act as a fluid).

Solar Atmosphere

The Sun's gaseous sphere consists of a number of different zones or shells. Below the visible photosphere is a convection zone through which the inner radiative processes give off energy to the surface. Overlying the photosphere is a layer known as the chromosphere with a depth of about 10,000 km (6200 miles) and a temperature of some 50,000 °C. The outermost zone is the corona, and in parts of it the temperature may rise as high as 1,000,000 °C. However, its density is very low; it has no well-defined limits and really represents the transitional zone between the inner gas shells and interplanetary space. Except during brief intervals during a total eclipse of the Sun, neither the chromosphere nor the corona can ordinarily be seen, but with instruments such as the chronograph – using monochromatic light – they can be observed at any time the Sun is above the horizon. At the time of total solar eclipse, the corona is visible as a beautiful pearly halo surrounding the Sun, and the shape it takes has a definite bearing on the age of the sunspot cycle.

47

Solar Activity

When the photosphere is observed telescopically, in addition to sunspots, a kind of granulation can also be seen. This is owing to the separate appearances of countless independent uprisings of convected gases from the interior which have lifetimes of about three minutes' duration. This activity gives rise to an ever-changing mottled pattern on the Sun's surface.

Another visible feature are the *faculae* – Latin for 'flames'. These are bright areas or filaments about one-tenth brighter than the surrounding photosphere. They probably represent a hotter layer immediately below the surface which swells up. Faculae are best observed near the limb of the Sun, since, owing to an effect known as limb darkening, the lower surface region appears less bright. They are connected in some way with sunspots, for often they first appear in a position which subsequently develops into a dark spot. In monochromatic light the faculae are always visible in the chromospheric region directly above the ordinary visible forms which lie on the lower photosphere.

By far the most impressive activity concerns the dark-red clouds of hot gas called *prominences* which are often seen projecting beyond the edges of the Moon at the time of total solar eclipses (Fig. 47). Prominences are composed of hydrogen gas, and with the use of a suitable filter they can be observed outside the time of eclipses. Often geyser-like eruptions and huge arched masses, moving at speeds of over 160 km (approx. 100 miles) per second, rise to great heights above the Sun's surface layer, and they have been successfully photographed with time sequence ciné-photography. Smaller prominences called *spicules* are also commonly observed near the Sun's polar regions, which, travelling at speeds of between 20 and 50 km (12 to 30 miles) per second, often pass into the outer coronal regions.

Solar Flares

The solar flare is the brightest eruptive activity on the Sun. It is a sudden, short-lived increase in the light intensity of the chromosphere in the region of a sunspot. The brightest examples can be seen directly in white light, but the average flare is normally visible only in hydrogen (Hα) light through a spectrohelioscope. They are very important events as far as the Earth is concerned, for they give rise to intense short-wave X-rays, cosmic rays, and corpuscular radiation which, if the orientation is correct, reaches the Earth some 26 hours later. This radiation interacts with the Earth's magnetic field and unbalances the ionospheric region which normally reflects back radio waves transmitted by broadcasting stations. Radio blackouts, sometimes lasting many days, are the result plus brilliant auroral displays in the polar and temperate regions (*see* p. 27). Magnetic compasses, also may be temporarily put out of action, or they may swing in such ways that they are of little use as navigational instruments.

Nowadays solar flares are of especial importance, since the high-intensity radiation they produce is harmful to astronauts outside the protective shield of the atmosphere. Monitoring stations spread round the world constantly keep watch for signs of solar-flare activity. The visible light effects can be seen on Earth about eight minutes after the event takes place on the Sun (since light travels at a speed of 299,460 km [186,000 miles] per second).

During World War II, radio waves originating from the Sun were detected for the first time in the military radar tracking stations. There is a definite link between radio waves and solar-flare activity. However, a flux of radio waves can be detected at all times and probably originates from the coronal region.

The Solar Wind

During the past decade the phenomenon called solar wind has been much discussed. Actually the so-called wind consists of a steady stream of charged particles, protons and electrons emitted by the Sun into interplanetary space at speeds attaining 600 km (370 miles) per second. The wind can be described as a moving cloud of solar plasma. It can be detected by reference to the observed changes in the Earth's magnetic field, by space probes and by the tails of comets. Further out in interplanetary space, nearer the orbit of Mars, the solar wind becomes a 'solar breeze', having by this distance lost much of its initial intensity.

The Solar Constant

The solar constant defines the quantity of heat received per minute by 1 cm^2 of the Earth's surface exposed to the vertical rays of the Sun, and equals 2.0 calories after corrections have been made for atmospheric absorption. The value of this constant has an important bearing on all life on Earth. Although at the present time it appears constant, in the past it may have been variable, and from time to time different theories are put forward that minor fluctuations may well be the cause of the Ice Ages. In 1967 the solar constant was measured by high-flying X-15 aircraft, and a value of about 2.5 per cent less than the best value commonly assumed was found. This result gave a figure of 1⅓ kW of energy for each square metre (approx 1.2 square yards) of the Earth's surface.

Solar Chemistry

To deduce the chemical make-up of the Sun, it is necessary to analyse the emitted light through the spectroscope. After the early discovery that light could be split into its component colours with a prism, the English astronomer Wollaston in 1802, and later the German astronomer Fraunhofer in 1814, found that if solar light was first passed through a narrow slit, a pattern of dark lines appeared superimposed on the spectrum. These lines are now called the Fraunhofer lines, and later in 1859, G. R. Kirchhoff and R. W. Bunsen (of Bunsen-burner fame) identified them with lines emitted from known chemical substances on Earth. The lines appear dark because the Sun is surrounded by absorbing layers of cooler gases. The cooler gases absorb the light emitted by the hotter layers below and are characteristic of the different elements contained in the outermost layer of the Sun. Nowadays we know of over seventy elements which occur in the Sun.

The Moon

After the Sun, the Moon is the most influential body which acts on the Earth, and its effects can be experienced directly on day-to-day events. In the past these events were attributed to the Moon's mystical qualities, and it is still fashionable in astrology and romantic literature to take these qualities into account. Nowadays, however, our chief concern is with the physical influences exerted by the Moon which are based on the more concrete fundamental laws of physics and dynamics.

The Moon is the Earth's natural satellite and nearest permanent celestial neighbour. It lies at a distance of 384,000 km (238,000 miles), and although small asteroids often probably pass at a much closer distance, their orbits are focused at the Sun's centre and are measured in periods of years, while the Moon revolves round the Earth in a period of 29 days 12 hrs 44 mins (synodic period). The *synodic* month equals the calendar month and represents the Moon's revolution once in respect to the Sun. The length between two identical phases= 1 lunation period= 1 synodic month. However, since the Earth moves 30° forward in its own orbit round the Sun during this time, the *sidereal* month (in respect to the stars) is about 2 days shorter= 27 days 7 hrs 43 mins.

Phases of the Moon

The Moon revolves round the Earth in an anti-clockwise or counterclockwise direction. The orbital movement changes the visible portion of its surface in an ever-repeating pattern. These are the phases of the Moon which complete one cycle over its month-long orbital period. New Moon occurs when the Moon is positioned exactly in line between the Earth and the Sun so that none of its illuminated surface is visible from the Earth (Fig. 14a). This definition of new Moon is slightly different from the new Moon of the ancient calendar makers who considered that the occurrence

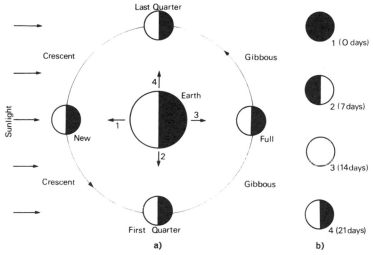

Fig. 14 (*a*) **Phases of the Moon as seen from the Earth during one synodic month.**
(*b*) **Age of the Moon in days according to phase.**

took place at the first sign of a slender crescent. The age of the Moon is usually expressed in days that have elapsed since the previous new Moon (Fig. 14b).

After the occurrence of new Moon, when it appears in the western sky shortly after sunset in its slender crescent form, the darker portion of the disc is faintly lit by Earth-light. This feature is sometimes called Earth-shine and is caused by sunlight reflected back off the Earth. It can be seen in some degree until shortly before full Moon and again shortly after. First Quarter occurs when half the Moon's visible surface is illuminated, Full Moon when the whole disc is illuminated, and Last Quarter when again only half the Moon's surface is illuminated (Fig. 14a).

The Metonic Cycle

In the fifth century B.C., the Greek, Meton, discovered what is now called the *Lunar Cycle of Meton*, or simply the Metonic Cycle. Meton found that after a lapse of 19 years, the phases of the Moon recurred on the same days of the same months (within about 2 hours). The number 19 is the smallest number of years that is a multiple of the *synodic* month = one lunation period, and there are very nearly 235 synodic months in 19 Julian years.

If the dates of full Moon are recorded during one 19-year cycle, they are therefore known for the following cycles. The cycle was of great significance to the old calendar makers, and the lunar phases shown in the *Book of Hours* (Fig. 41; p. 30) were all obtained by reference to the Metonic Cycle. The number 19 became known in the Middle Ages as the *Golden Number*, for the significant dates were inscribed in gold upon public monuments, and as one contemporary quoted: 'This number excels all other lunar ratios as gold excels all other metals.'

Moon's Rotation Period

The fact that the Moon always turns the same face towards the Earth implies that during one orbital *revolution* round the Earth it also *rotates* once on its axis. This is owing to a tidal coupling between the Earth-Moon system. However, due to the Moon's elliptical-shaped orbit and its axis of rotation not being quite perpendicular to the plane of the orbit, a rocking or oscillating motion is induced which is termed *libration*. This has the periodic effect of displacing the visible hemisphere of the Moon so that we are able to see four-sevenths of its surface area from the Earth.

Size and Appearance

By a remarkable astronomical coincidence the apparent size of the full Moon equals that of the Sun. But in comparison with the Sun, the Moon is a much smaller body, and its apparent equal diameter is only because it is located much closer to the Earth. Even by comparison with our own world, the Moon is a much smaller body and occupies a volume fifty times less (Fig. 35). The Moon's diameter is 3476 km (2159 miles), and it is slightly flattened at the poles.

Most people have at some time remarked on the phenomenon that the Moon appears larger when near the horizon (actually it is slightly smaller owing to the effects of refraction). This appearance is known as 'the Moon illusion' and has been explained by psychologists as being due to properties of the human eye and brain. It is brought about by the fact that the eye relates the size of an object to different surroundings. In the wide expanse of mid-heavens, there is nothing with which to compare it, but near the horizon there are many familiar objects.

Tides

The gravitational forces exerted between the Earth and the Moon are the direct cause of the tides which occur throughout the oceans and seas of the Earth and to a lesser extent over the land masses. The proper understanding of the cause of tides was one of the first direct results of Newton's work on gravitation. The Sun, however, is also an important influence in raising tides, and exerts a force equal to about 30 per cent of the pull from the Moon.

Although the underlying ideas about tides can be explained in simple terms, some of the minor, irregular influences are of great complexity. If the Earth were a wholly rigid body, *all* the tidal effects would occur in the oceans and seas. If the Earth were a perfectly elastic body, with no rigidity, the water tides would hardly be perceptible.

It is fairly easy to predict how tides should behave, or rather the total force that the tides should exert on the oceans and seas. The effects of currents and shore lines on the whole available force is difficult to calculate accurately, but in round figures only 70 per cent of the available force can be accounted for in the oceans. This means that the remaining 30 per cent is taken up by the Earth's crust and is a direct indication of the elastic properties of our planet.

Most of the Earth's oceans and seas experience two *high* tides and two *low* tides in the period of 24 hrs 50 mins. Although the Moon during this period has crossed each meridian only once, the reason for *two* tides is that the pull of the Sun and Moon distort the oceans, both in the direction *towards* and *away*, so in effect a bulge of water is created on both sides of the Earth (Fig. 15). The height of tides varies quite widely. Its amplitude may range from only a few centimetres in the waters of small lakes to over 12 m (40 ft) in the Bay of Fundy. The amplitude depends very much on the position of the Moon in its orbit. At the time of the New or Full Moon, when the Sun and Moon are lined up to produce a combined gravitational pull, the tidal amplitude is greatest. This is the period of *spring* tides when the high and low tides occur. At the time of First or Last Quarters, the pull of the Sun and Moon work against each other, and the resulting low tides are called *neap* tides (Fig. 15b).

Successive high tides at any one place on the Earth (which on average occur once every 12 hrs 25 mins) are never the same amplitude, owing to the Earth's axis being tilted at an angle.

Additional tidal effects worth noting are:

(a) At places near the coast, the time of high water lags some 6 hrs behind after the crossing by the Moon.

(b) Land tides quickly build up to the maximum peak and then quickly subside since the crust has great rigidity.

(c) The friction of the tide robs the Earth of one ten-billionth part of its rotational energy.

(d) Tides occur on the Moon of greater amplitude than on the Earth and indeed may produce certain observable effects in various surface features.

(e) As a consequence of tidal forces, very slow changes occur in the physical relationships of the Earth-Moon system. These changes affect the length of day and month and also the distance separating the two bodies. The energy of spin – or angular momentum – of the Earth cannot be destroyed, but the result is an acceleration in the orbit shape and size. Sometime, in the far distant future, the effect may produce a lunar month of 50 days. Still further ahead, the Moon may become totally destroyed by tidal forces and end up as a ring of small bodies encircling the Earth similar to the Rings of Saturn.

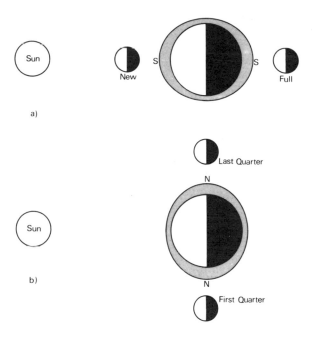

Fig. 15 The tidal bulge and spring and neap tides:
(a) Spring tides (S) – greatest amplitude – occur when the Sun and Moon are lined up at the time of New and Full Moon.
(b) Neap tides (N) – lowest amplitude – occur when the Moon is at First or Last Quarter.

Interior of the Moon

In November 1969, the Apollo 12 astronauts set up the first automated geophysical laboratory on the *mare* known as Oceanus Procellarum (Fig. 62). Since that time and the later Apollo flights, the seismometers have transmitted back to Earth a complete record of all moonquakes which have occurred.

On average we know of one moonquake per day, but these are small affairs in comparison to the daily tremors recorded in the Earth's crust. Once a month, however, at a precise 28·4-day regularity, much larger moonquakes take place which originate from many sites near the seismometers. These quakes occur at the time when the Moon makes its closest approach to the Earth (at perigee), and are simply the results of a bulging effect which raises a lunar land tide about 1 m (1·09 yds) in height, but which has considerable influence to a depth of several kilometres.

The recorded signals received from the moonquakes have provided a great deal of evidence about the *probable* interior of the Moon. The Moon's crust appears to be more rigid than that of the Earth. Neither does it seem to have a layered structure like the Earth and possibly contains only two kinds of primitive rock, which *may be* in solid state even at the very centre. Other indications are that there is no free water inside the Moon, and as a consequence the inner rocks are welded much more tightly than those in the interior of the Earth.

In spite of these findings there are alternative ideas about the Moon's interior. One theory of the seismic wave patterns is that there must be a 6 km (3·7 miles) deep, highly *compacted* layer of dust overlying regions of the Moon which has the consistency of a monolithic mass of cemented rock. Such a layer would account quite satisfactorily for the data so far received, but at present this idea is not widely accepted. Another idea supports a suggestion of interior convection currents as in the Earth's mantle, although estimates arrived at from electrical conductivity experiments have indicated internal temperatures of the range 800 to 1000 °C. These temperatures are certainly too low to allow convection as we visualize it in the Earth's interior. However, the same electrical experiments have also shown the existence of a weak magnetic field whose cause could certainly be attributed to internal convective forces.

Nature of the Lunar Surface

Although samples of the lunar surface have now been brought back to Earth, the problem regarding the origins of the various features on the lunar surface is still far from being solved. If anything, the problem has become infinitely more complex! Prior to the first manned and unmanned landings, selenologists were broadly divided into two main camps: those who favoured a volcanic origin for the craters, and those who favoured an impact (meteoric) origin.

At the present time the consensus of opinion believes that *both* agencies have contributed. We have conclusive evidence in the form of rock samples to show the volcanic nature of certain features and likewise fairly conclusive evidence that many of the sharply defined craters have been caused by impacts of bodies with the lunar surface. The problem is complicated by the fact that the volcanic lava flows *may* have been triggered off by severe meteoric impacts.

The so-called highland areas (lighter coloured features) have a density of 2·9 and may be floating on basaltic gabbro of 3·3 – in a similar fashion that the lighter rocks appear to float on the denser rocks in the Earth's crust. The Moon's density in the interior does not increase much, for its overall density is only 3·35.

Since the Moon has no atmosphere or moving water, weather plays no part in erosion or denudation of the surface. But erosion is certainly a significant feature as the rounded lunar rock samples conclusively show. Two other influences might play an important role. Since the Moon has no cushion of air to protect it, even the smallest meteoroids must reach the surface intact and have much of their cosmic velocity remaining as they impact on the exposed lunar rocks. Also the temperature has an extremely wide range between day and night, from 102 °C down to −157 °C. This must have some significance in shaping land forms, for even on Earth rapid cooling of rocks in desert regions induces contractions that cannot be contained and results in an exfoliation process which gives the rocks a rounded appearance. Another influence which is possibly less significant is brought about by the direct bombardment of the Moon's surface by solar flare radiation, an effect which is known as solar sputtering.

Mascons

A discovery of dense material below the surface of the Moon was one of the surprises of the Lunar Orbiter photographic flights, which surveyed the Moon in detail during the late 1960s.

The name Mascon is derived from the two words *mass concentration*. They were detected from the strong gravitational pulls they exerted on the orbits of the unmanned satellites. Satellites in orbit are extremely sensitive to anomalous gravitational forces, and it was noted that in the regions of the maria the orbits were distorted. These anomalies have been interpreted in two different ways. They could be due to large nickel-rich meteorites buried just below the lunar surface which give rise to isolated high-density spots.

Or they could have occurred during the formation of the Moon and are regions of abnormally dense material which became 'frozen' into the lunar crust. They are of particular interest in connection with the unexplained small oscillations which sometimes affect the Moon in its orbital motion that hitherto had no obvious explanation.

Principal Surface Features

Even to the unaided eye the Moon can be seen to have a variety of surface markings. The early astronomers who viewed it, concluded that the darker areas must have been seas similar to those on Earth. As a consequence they are still so called in lunar nomenclature and go by the name *maria* (Latin for 'seas') even though it has long been realized there is not a vestige of water present. When the Moon is full, the dark maria suggest the appearance – to the unaided eye – of the traditional 'Man in the Moon'. With a little imagination they also sometimes suggest the pictures known as 'Woman reading a book' or 'the Crab's claw'.

In addition to the dark maria, often referred to as the lunar plains, there is also a huge variety of craters and crater-like features which range in size from tiny pits to several miles. Some appear to be of more recent origin than others, and the ring plains are often barely discernible except under certain angles of illumination.

Mountain chains whose peaks often exceed 7000 m (20,000 ft) occur near the borders of the maria. They are particularly noticeable from the long dark shadows they cast when they are observed near the *terminator* – the name given to the dividing line between the daytime illuminated portion of the surface and the lunar night.

A more obscure feature of the surface is the bright lunar rays. These are best observed at the time of full Moon and measure up to 2500 km (1500 miles) in length. They appear to radiate from many of the larger craters (Fig. 62).

An interesting telescopic feature is the network of fine cracks, called rills or clefts, which are more noticeable near or on the maria. They often extend for many miles and again they are best seen near the terminator when their interiors are filled in dark shadow (Fig. 50). The maria also contain wrinkle-like ridges and features which appear to be geological faults.

Lunar Surface Rocks

The recovery of the first lunar samples was one of the most important events in the history of our knowledge of the Moon. To date, in the early 1970s, no organic material has yet been identified, but it must be said that none was expected except perhaps organic-like hydrocarbon material similar to that found in the fairly rare carbonaceous meteorites.

Many rock samples, compared with those found on Earth, contained surprisingly large amounts of titanium, zirconium and ytterbium. And they appear to be much more abundant elements on the Moon than in meteorites, yet concentrations of iron and magnesium are lower. Many of the specimens have been classified into rock types that resemble terrestrial volcanic lavas. One of the least expected results was that many samples show signs of erosion, for they are markedly rounded on their upper (exposed) surfaces. There is now a well-known description of the appearance of an eroded specimen when it was first picked up by the crew of the Apollo 11 Mission. The log reports that it appeared to be a partially buried, rounded object that looked exactly like a motor-car distributor cap.

One theory suggests that lunar rocks may be eroded by the surface impact of small meteorites, for the exposed samples are frequently pitted with holes ranging from the microscopic up to 2 mm in size. Many of these tiny pits are lined with glassy substances. The Apollo crews also reported much larger 'glazed' surfaces, and it has been suggested that their origin may be due to intense activity on the atmosphere-free lunar surface by radiation originating from solar flares. Other varieties of rock include loose aggregations like breccias.

Harvest Moon

The rising of the Moon normally occurs about 52 minutes later each day owing to its anti-clockwise orbital motion round the Earth. However, the daily intervals between successive risings can vary greatly and may be reduced to as little as 13 minutes, depending on the time of year and the location of the Moon in its orbit. At the time of the autumnal equinox (northern hemisphere), vernal equinox (southern hemisphere), the full Moon's declination is such that it gives rise to a short-period effect, for an interval of a few days, of appearing to rise about the same time soon after sunset. Since this time of year is the traditional harvest time in the northern hemisphere, the phenomenon has become known as the Harvest Moon.

Moon's Origin and Age

Since the first lunar samples were brought back to Earth, we know that the Moon is certainly no newcomer to the solar system. The lunar rocks, in common with meteorites, contain significant amounts of the radioactive isotope aluminium-26. This fact makes it implicit that the Moon has been exposed for a great length of time to bombardment by cosmic rays. Its age is also borne out by physical analysis which uses the potassium-argon technique to measure, in rock samples, the slow conversion of potassium into argon. Figures derived by this method have given the Moon's age as 4600 million years, which is equal to the Earth's estimated age. Some additional analysis of the concentrations of short-lived radioactive material, produced by cosmic rays, indicate that many of the recovered surface rocks have lain undisturbed for from between 20 million and 160 million years.

The question of the Moon's origin, however, has not yet been satisfactorily answered, and *all* ideas must be tentative ones. Theories range from its common origin as part of the Earth – and its breaking away in the dim past – to its capture by the Earth from a previous independent orbit round the Sun.

Sir George Darwin (1845 1912), son of Charles Darwin, calculated that the Moon was once in contact with the Earth, and at this time the two bodies had a rotation period of about 4 hours. They subsequently became separated because of resonant tide-raising effects. But there are many objections to this idea. Present-day evidence tends to favour the theory that both the Earth and the Moon were formed near each other by the process of material accretion. Some evidence suggests that they evolved as cool bodies which later heated up when trapped radioactive minerals began to decay in their interiors. The Moon's present surface is probably much more ancient than the Earth's present surface, and when lunar exploration has progressed further, it may well provide more definite clues to the origins of both bodies.

Lunar Influences

Apart from the dynamical (gravitational) effects on the Earth which gives rise to tides, there are some interesting biological ones. Nature can show many examples of life which has adapted to the light cycle of the Moon's phases or the tidal influences. One example of adaptation to lunar rhythm was first observed by Columbus in the Caribbean in 1492. On 11 November, he observed a mysterious light at sea, and when the ship drew closer, they saw the water swarming with glowing worms. Subsequently, the life cycle of these polychaete worms was discovered to be tied to the waxing of the Moon. When the Moon is three-quarter full, the female worms come to the surface at night and deposit a scintillating spawn to attract the male. Likewise the North Sea herring is greatly influenced by the presence of the full Moon; the human female cycle which averages 28 days must also certainly have had some evolutionary stimulus when our species was evolving.

Less scientific lunar influences are attributed to variation in the Earth's climate and weather. Even in the 1970s a well-known Soviet geophysicist reports in his findings that the Earth's temperature is affected by the Moon over the 19-year Metonic cycle (*see* p. 51). However, it is difficult to see what possible climatic effects the Moon could exert on the Earth unless atmospheric tides are a significant influence.

Eclipses

The Earth moves round the Sun once every year, and the Moon moves round the Earth once every lunar month. These are the essential facts which bring about both solar and lunar eclipses. The word eclipse is from the Greek *ekleipsis*, meaning forsaking, quitting or disappearance. A lunar eclipse occurs when the Earth's shadow intercepts the surface of the Moon. A solar eclipse occurs when the Moon passes between the Earth and the Sun.

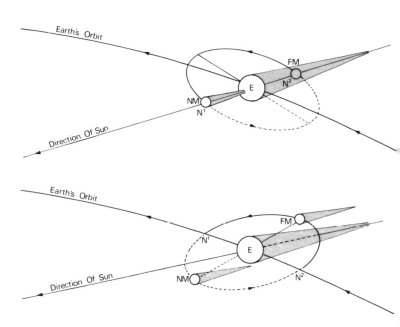

Fig. 16 Eclipse occurrences. Owing to the Moon's orbit being inclined a little over 5° eclipses can only occur at New Moon (NM) and Full Moon (FM) when the line of nodes (N¹–N²) of the Moon's orbit coincides with the direction of the Sun.

Were it not for the fact that the Moon's orbit does not lie in the same plane as the Earth's orbit, there would occur about 25 solar and lunar eclipses each year. The inclination of the Moon's orbit is a little over 5°, so that eclipses can only happen at the time when the nodes of the Moon's orbit coincide with the direction of the Sun (Fig. 16). The two nodes (or crossing points) are termed *ascending* when the Moon moves from south to north, and *descending* when the Moon moves from north to south. This gives rise to eclipse periods when eclipses may occur during the year because the Sun-Earth-Moon geometry is suitable.* Owing to these geometrical configurations only a maximum of seven eclipses of either kind can occur within one year and sometimes it may be as few as two. In the years when there are seven eclipses, five of these may be of the Sun and two of the Moon. When there are only two eclipses, both of these must be of the Sun. There can never be more than three eclipses of the Moon, and in some years there will be none at all.

It is very rare to have five solar eclipses in one year. Eclipses of the Sun are more numerous than those of the Moon in the proportion of about 3 to 2, yet at any given place on the Earth, more lunar eclipses are visible than solar eclipses, simply because the former, when they occur, are visible over the whole hemisphere of the Earth which is turned towards the Moon, while the area over which a total eclipse of the Sun can be seen is only a narrow belt 240 to 275 km (150 to 170 miles) in width. However, *partial* eclipses of the Sun are visible over a much wider area on either side of the path travelled by the Moon's shadow.

The Saros

Babylonian astronomers observed that eclipses repeat themselves in a cycle over 6585·32 days (18 years 11⅓ days). Thus, if an eclipse occurred today about noon, after 18 years 11⅓ days another eclipse would occur, although not exactly in the same geographical location. This is termed the *Saros* and enables one to compute and predict eclipses with fine accuracy. However, if greater accuracy is required, three Saroses give better results, and for even greater accuracy one needs to take into account 48 Saroses.

The Saros is brought about by the fact that the nodes of the Moon's orbit do not maintain fixed positions but move backwards, or retrograde, about 19¼° each year, so that they make one complete revolutionary cycle in 18·5997 years.

---

* Called syzygy = three astronomical bodies in a straight line.

Solar Eclipses

Eclipses of the Sun are termed total, annular or partial (Fig. 17). A total eclipse occurs when the Moon covers the entire solar disc, and this gives rise to some highly spectacular phenomena (*see below*). An annular eclipse takes place when the Moon, owing to its eccentric orbit, does not cover the entire solar disc, and there remains a small circle of Sun, or an annulus, surrounding it. A partial eclipse occurs when only part of the Sun's disc is obscured by the Moon. A partial eclipse will always be visible on either side of the track of the Moon's shadow at the time of a total solar eclipse.

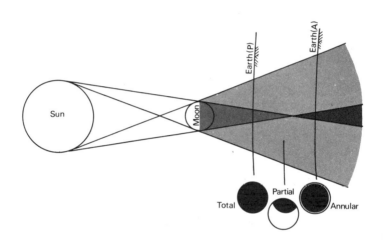

Fig. 17 Solar eclipses: total, partial and annular. The schematic representation shows the Earth's surface at the time when the Moon is furthest away from the Earth in its orbit: at apogee (A), and nearest at perigee (P).

Total Solar Eclipse

The most impressive of all the eclipse events from a spectator's point of view is a total eclipse of the Sun. In early times the occurrence of a total eclipse often had wide historical repercussions if we are to believe the ancient chroniclers. Wars were either won or lost depending on what significance the events had on the combatants who were fighting at the time the eclipse occurred.

There is the well-known apocryphal story concerning the two Chinese astronomers Ho and Hi who were put to death by the emperor Hsia Chung-K'ang for not predicting a total eclipse of the Sun which supposedly took place in 2136 B.C. They are remembered by the anonymous verse:

> *Here lie the bodies of Ho and Hi,*
> *Whose fate though sad was visible –*
> *Being hanged because they could not spy*
> *Th' eclipse which was invisible.*

The observation of a total eclipse of the Sun is a most memorable event and leaves a lasting impression on the human mind. After first contact, when the Moon begins to creep across and obscure the face of the Sun, the intensity of daylight is slowly diminished. The air temperature also drops, and as darkness approaches, birds fly to roost in their nests, imagining that nightfall is about to overtake them. The landscape is cast with an eerie gloom, and as the time draws near, the trees and grass lose their familiar colouring. Just before totality (second contact) one of the most impressive features is the appearance in the distance of the fast-moving shadow of the Moon which finally engulfs the observer at a speed of 2940 km (1830 miles) per hour. Sometimes, peculiar shadow bands can be seen, which are owing to a refractive effect in the Earth's atmosphere. As the Sun is completely obscured and darkness falls, the corona suddenly bursts into view and creates a pearly-white halo which surrounds the Sun and varies in shape depending on the particular year of the sunspot cycle. Visible against the white lustrous background are the red or purplish-red prominences which look exactly like huge red flames (Fig. 39).

Throughout the totality phase the sky remains dark, and many of the brighter stars and planets can be easily picked up. But the sky is not completely pitch-black as many suppose; it is rather more like a late evening twilight effect. At third contact the Sun re-emerges as suddenly as it was previously engulfed, and within seconds the sky has brightened, and the birds have reawakened to the second dawn of the day. Minutes later, the chill eclipse air has already warmed to the touch of the Sun.

Another phenomenon, which is often noted, goes by the name of Baily's Beads and occurs immediately before second contact. It is so named after the Englishman who first discovered it in 1836, but if justice were done, they should be called 'William's Beads' after the American observer who

63

saw them at least four decades before Baily gave his account. The 'beads' are produced by the dying rays of the Sun shining through the gaps remaining at the uneven edge of the Moon just before it finally engulfs the Sun. They recur prior to the third contact when they appear to run one into the other like so many drops of water. The highly descriptive diamond-ring effect is also a prominent feature of solar eclipses. It occurs before Baily's Beads at second contact and precedes it again before third contact, when the sparkling diamond effect is even more striking (Fig. 55).

Fig. 18 The telescopic appearance of Baily's Beads (*see also* Fig. 139).

Total solar eclipses offer unique opportunities for many programmes of work. Sometimes comets are discovered near the Sun, but generally these eclipses are significant in providing astrophysicists an opportunity to study various levels in the solar atmosphere. They also allow a practical test in connection with Einstein's theory of relativity. One facet of this theory is that light has gravitational mass and, therefore, if it passes close by a larger body it must be deflected. At the time of total eclipse, stars on either side of the Sun can be photographed. If the theory is correct, the apparent positions of these stars will be shifted in relation to their apparent positions in the night sky when the Sun is removed 180°. Although the predicted deflection is small, it nevertheless should be measurable. First attempts to verify it were made in May 1919 by Sir Arthur Eddington, the British astronomer, who later popularized the idea of the expanding universe. This result and subsequent results showed that indeed a shift was measurable. However, it would appear to be larger than that predicted by the theory, and there may be other factors (at present unknown) which also have some significant effect.

The longest possible duration of a total solar eclipse is a little under 8 minutes, and the most favourable locality is at about latitude 20°N when the Sun is near apogee, but usually the time of a total eclipse is much shorter.* An annular eclipse of the Sun may last a little over 12 minutes.

* For example, the total eclipse of the Sun observed by the author in Siberia, 22 September 1968, lasted only 38 seconds.

Lunar Eclipses

In many respects lunar eclipses are also impressive events and again they had deep significance in the ancient world. It is related that the lunar eclipse of 1 March 1504 saved the life of Christopher Columbus. He was threatened with death by starvation in Jamaica where he was refused provisions by the locals. Forewarned of the arrival of this eclipse by his almanac, he threatened to deprive them of the light of the Moon and of course kept his word. When the eclipse began, the terrified natives flung themselves at his mercy and subsequently brought all the provisions he required.

Lunar eclipses are influenced by the two kinds of shadow cast by the Earth. In Fig. 19 it will be noted that the darker shadow called *umbra* has a narrower pencil-like beam than that of the outer shadow called *penumbra*. When the Moon is near one of its nodes, the umbra shadow of the Earth will cover its face completely, producing a total lunar eclipse. Further from the node point, only part of the umbra will fall on the Moon and produce a partial eclipse. Yet further away from the node, only the lighter penumbra will sweep the Moon; beyond 10° from the node point there will be no eclipse.

Lunar eclipses last for much longer periods than total solar eclipses. Maximum time for passing through the umbra region is 2 hours, and 4 hours for passing through both the umbra and penumbra.

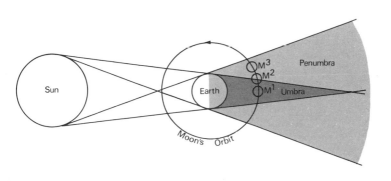

Fig. 19 The umbra and penumbra shadows. A total eclipse of the Moon occurs at M¹, a partial eclipse at M², and a penumbral eclipse at M³.

Notable colour effects accompany lunar eclipses, which vary considerably depending on atmospheric conditions. At the time of mid-eclipse there is a distinct coppery glow from a much-dimmed Moon. The Moon itself rarely disappears completely as it should when obscured by the umbra. This is due to light reaching it via the Earth's atmosphere which refracts or bends light round the edges, and some reaches the Moon's surface. The coppery tint is thus caused by the absorption of the blue end, or short wavelengths in sunlight (similar to the colour effect of the setting Sun on Earth).

At times when the Earth's atmosphere is polluted with suspended volcanic dust, little light reaches the lunar surface via the atmosphere, and it may seem to the naked eye to disappear. This happened in 1964 after the volcanic activity in the East Indies when the eclipse which occurred on 19 December was one of the darkest on record. The Moon could not be seen at all with the naked eye, but remained just visible in 10×80 binoculars.

Mercury

Mercury revolves round the Sun in a period of 88 days. It is one of the smallest planets, with a diameter of 4840 km (3005 miles), and its volume is only 0·06 that of the Earth. It is also the closest known planet to the Sun, which makes it a difficult object to observe with the naked eye except for short periods just before sunrise or sunset, at the time of greatest elongation in its orbit. Legend tells us that it was never seen by Copernicus owing to the prevalence of misty horizons where he lived near the banks of the river Vistula in Poland. However, it was certainly known as far back as 264 B.C., when it was observed by the Babylonians, and the Greeks called it *Stilbon*, the 'twinkler'.

Until very recent times optical observations were interpreted to indicate a rotation period equal to the revolution period analogous to the situation with the Moon. But radar measurements indicate a period of 59 days, or nearly two-thirds the period of revolution. This is significant as it suggests that the period is no casual value, but rather a result produced by a resonance effect with the tidal forces of the Sun. The consequences of this resonance period are interesting for it means that Mercury's total day is three times its sidereal rotation period; the planet then must present the same surface to the Sun during the course of two of its years (or 176 Earth days).

Mercury has a highly eccentric orbit which takes it to within 45·6 million km (28·3 million miles) of the Sun at perihelion and 69·6 million km (43·2 million miles) at aphelion. The plane of the orbit is inclined at 7° to the ecliptic and is the steepest inclination for any of the principal planets excepting Pluto.

Infrared studies of Mercury's surface reveal that the temperature drops from a maximum of 350 °C at noon to −160 °C at midnight during the 59 Earth-day transition. These studies also show that the surface must be made up of the same kind of rock as the Moon. But the surface temperature of Mercury must rise considerably higher than that of the Moon owing to its closer proximity to the Sun. Any atmosphere which exists must be extremely tenuous, for the velocity of escape from the surface is only 4·2 km (2·6 miles) per second, compared with 11 km (7 miles) for the Earth, and bombardment from solar flares would tend to 'blow' away all but the heavier surface gases.

Optical observations over a long period of time support the recent work concerning the probable surface features and Mercury's lack of atmosphere. It is not surprising that the optical astronomers were mistaken about the planet's rotation period, for the illuminated surface features are difficult to observe. Best times for observation are during daylight hours when Mercury is high in the sky and removed from the atmospheric turbulence (or poor 'seeing') which occurs near the horizon. When 'seeing' is at its best, vague darkish markings are often distinctly noted, and more rarely hazy-white obscurations which some observers claim to be dust storms.

The reflecting power, or albedo (from the Latin *albus*, 'white'), of the surface features gives a value of 0·06, which is quite low, but again this is another indication of its possible similarities with the lunar surface (0·07).

Naked-eye observation is best attempted when Mercury is a morning 'star' and rises just before the Sun in September or October at a time coincident with its maximum westward elongation. At this time it will shine as a bright distinctive star of mag −1·8.

There has been much speculation about the interior of Mercury. Its density is about 5·5 and may contain a considerable iron core or some other heavy elements. The fact that the planet is nearest the Sun, and also one of the most dense bodies in the solar system, is significant in the understanding about the origin of the planetary system, for as we recede from the Sun to the outer, giant planets, we find much lighter bodies. Further knowledge of Mercury awaits future space missions.

Venus

Venus is by far the brightest of all the planets, and near eastern elongation it can be picked up with the unaided eye even before the Sun has set. At this time it can also be seen with the unaided eye at midday if one knows *exactly* where to look. After dark, during its brightest phase, it will also cast a perceptible shadow.

Following Mercury, it is the next planet in order out from the Sun at a distance of 108·29 million km (67·26 million miles). It has one of the least eccentric orbits of all the planets (0·007) so that its perihelion and aphelion distances are similar. Like Mercury it also exhibits a phase effect as it revolves round the Sun in a sidereal period of 224·7 days in a path inclined to the ecliptic at 3° 24'.

The rotation period of Venus is a problem which long perplexed the observational astronomers and estimates ranged from 24 hours to one Venusian year. The problem was a difficult one since the surface of the planet is perpetually enshrouded in a dense opaque cloud which gives the disc the brightest reflecting surface in the entire planetary system (albedo 0·75, compared with 0·07 for the Moon). Markings, if they exist at all, are most likely properties of the atmosphere. It was a great surprise then when the first radar results, obtained in the early 1960s, gave a rotation period exceeding the Venusian year. It was found that the planet rotated in a retrograde direction (clockwise) in a period of 243 days with its polar axis inclined at 3°. This means that its daily cycle is 117 Earth days and in three rotations it resonates with two *revolutions* of the Earth round the Sun.

One theory to account for the extremely slow rotation period, which is the slowest for a planet in the entire solar system, puts forward the idea that Venus once had a satellite. At an early period of history Venus probably rotated at about the same rate as the Earth. The satellite, however, about twice the size of the Moon, was captured from a retrograde orbit round the Sun. This resulted in robbing Venus of its spin energy (angular momentum). It also resulted in the satellite slowly spiralling inwards until it finally crashed into the planet. An extension of this theory also goes on to explain that the physical conditions now prevailing on the surface of Venus are a direct result of the frictional heat caused by the colossal impact of the satellite. At the present time this theory is *pure speculation*, and the reason for the slow rotation may be caused by a factor yet unknown.

Venus has often been called the sister planet of the Earth, for it resembles it both in size (12,390 km [7700 miles]) and mass (82 per cent). Nevertheless, the environment prevailing on the surface of Venus and in its atmosphere is very different from our own. The surface temperature is over 450 °C which is about the temperature found 20 miles below the surface of the Earth. One would have to descend to a depth of 400 km (250 miles) on both planets before the temperatures were equalized at a figure of about 1250 °C. It has been calculated that if the Earth were as hot as Venus on its surface, the oceans would evaporate to create an atmosphere so thick and heavy that pressure at ground level would be 300 times greater than at present.

The atmosphere of Venus is made up of between 75 to 90 per cent carbon dioxide by volume. Nitrogen plus the noble gases in the lower regions account for less than 5 per cent, and it is doubtful whether oxygen exceeds 0·4 per cent. Estimates about water vapour are difficult to assess at the present time but the figure must be very low. The depth of the Venusian atmosphere is much greater than our own, and the pressure it exerts is equivalent to about 100 times greater. A direct indication of this pressure is that all the early Soviet space probes to the planet were crushed before reaching the surface. However, Venera 7 did successfully transmit back signals after coming to rest on the surface.

Radar observations through the 21 km thick cloak of clouds have shown that the planet's equator is peppered with several large crater like features— some over 300 km in diameter—similar to those of Mars and the Moon. Through optical telescopes* the surface shows only extremely vague darkish patches, which may be due simply to effects of contrast. Occasionally a faint diffuse illumination is visible over the dark night side when the planet is in crescent phase. This has been termed the Ashen light, and a popular theory attributes its presence to electrical activity in the Venusian upper atmosphere – possibly akin to auroral activity on the Earth. However, present knowledge of the planet's magnetic field leans against this particular idea. Sometimes a blunting of the 'horn' cusps has been noted, often in small telescopes, and observed dichotomy (half-phase) does not coincide exactly with the calculated predicted time. Some optical observations made in the near ultraviolet region of the spectrum, show cloud-like formations in the atmosphere which appear to circle the planet in a retrograde motion in about a 4-day period.

Venus has little or no magnetic field, and this is consistent with its slow rotation period. There would seem to be no equivalent of the Earth's Van Allen belts surrounding the planet, so that the influence of the solar wind must be much less marked. Since there is little water and extremely high temperatures, the surface must be very hostile to any kinds of life form.

As a naked-eye object there is no mistaking its presence in the sky. It is not surprising that in the guises of Hesperus and Phosphorus it became famous in the ancients' world as the morning and evening 'stars' before it was realized that it was the same object at different elongations. It becomes its brightest, shining at mag −4·4 (compared with Mars −2·8, Mercury −1·8), 35 days *before* western elongation and 35 days *after* eastern elongation, and the phase effect is easily observable in small binoculars (e.g. 8 × 30s). At the time of inferior conjunction it approaches the Earth within

* Observations best carried out during daylight hours.

38·6 million km (24 million miles) – which is 16 million km (10 million miles) closer than Mars approaches us at opposition.

At the inferior conjunction, if the orbital geometry is suitable, Venus may transit the Sun. However, this is among the rarer periodic events, and very few living persons can have witnessed one. During the present era the transits occur in pairs spaced eight years apart. The last pair took place in 1874 and 1882, and the next pair will not take place until 2004 and 2012. It is of interest that it was a transit of Venus which first took Captain Cook to the South Seas in 1769 to carry out observations from Tahiti.

Mars

Mars has long been known as the 'Red planet' and was the striking celestial representative of the mythical Roman 'God of War' because of its distinctive brilliant red tinge. At an earlier period, the Egyptians called Mars 'Horus-the-red', and it is the only Egyptian planetary name which is immediately identifiable. However, to most naked-eye observers its colour is a brick, or terracotta shade of red rather than a deep ruby, but when close to the horizon, it will often flash through a whole spectrum of false colours owing to the effects of atmospheric scintillation.

Mars has long fascinated mankind because of the possibility of finding some kind of life there. Since shortly after the invention of the telescope, prominent markings have been noted, and from the seventeenth century onwards, many generations of observers have carefully followed and mapped them. Nevertheless, it is only in the last decade that it has been known with any certainty what kind of conditions prevail on the planet.

Mars is smaller than the Earth with a diameter of 6800 km (4200 miles). It has a slight oblateness, but the actual value of this is still in dispute. Telescopic observations indicate the equatorial diameter to be 40 km (25 miles) greater than the polar diameter, but measurements derived on dynamical grounds, estimated by observations of the two tiny satellites of Mars, suggest a smaller figure of 17 km (10·5 miles). The rotation period is 24 hrs 37 mins 23 secs and its axis is tilted 23° 59′ to the plane of the orbit – both quantities remarkably close to those of the Earth.

The revolution of Mars takes 1·88 years or 687 days (sidereal) and the planet lies at a mean distance from the Sun of 228 million km (142 million miles). The path of Mars is distinctly elliptical. The eccentricity of 0·093 is large for a planet and this has its effects on observations made from the Earth. Because of its apparent closeness in space and the length of its orbital period, it only approaches an Earth opposition about once every 2 years. Really close approaches are influenced by the orbital eccentricity and occur

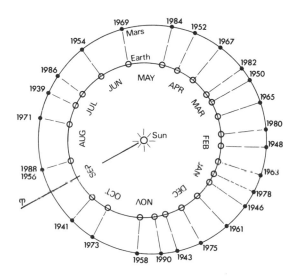

Fig. 20 Oppositions of Mars as seen from Earth 1937 to 1990. Opposition occurs about once every two years, but the closest oppositions occur at about 15/17-year intervals.

at intervals of about 15 years, or every seven synodical revolutions of Mars. During its orbit round the Sun, Mars, as seen from Earth, appears to advance in the sky across the backcloth of stars for 710 days and then make an apparent backward (retrograde) loop for 70 days. This is due to the Earth's own orbital movement round the Sun (Fig. 10c). Thus, as seen from Earth, Mars takes 780 days to reappear at the same position in the sky. This is called the synodic period as distinct from the sidereal period.

Because of the orbital eccentricity, opposition distances of Mars vary between a best figure of 56 million km (35 million miles) and a worst of 98 million km (61 million miles). It will be noted that the smallest distance occurs at 15/17-year intervals and coincides exactly with the times of Mars' perihelion and the Earth's aphelion (Fig. 20).

Between opposition and its two positions of quadrature, Mars appears gibbous and looks like the Moon three days before, or after, full. Galileo first detected the Martian phase with his primitive telescope in 1610, and writing to a friend he says: 'I dare not affirm that I can observe the phases of Mars, however, if I mistake not, I think I already perceive that he is not perfectly round.'

The prominent dark markings and the polar caps were also noted soon after the invention of the telescope. During the nineteenth century, the darker tints were thought to be seas like the darker regions on the Moon and likewise were called *maria*. Seasonal changes were noted, such as the expansion and shrinking of the opposed polar caps. Whitish transitory spots which resemble clouds were also often observed as well as unexplained partial obscurations of the darker tints.

In 1877, Mars made a particularly close approach to Earth, and the Italian astronomer Schiaparelli afterwards announced he had detected a network of fine lines criss-crossing its surface, and they were connected by small, round darkish spots. Unfortunately, he called the lines *canali*, apparently without intent of implying an artificial origin for them. However, the observations triggered off one of the greatest of all astronomical controversies which was not resolved until the middle 1960s. Astronomers became divided into two camps: the 'pro-canal' observers who claimed to be able to see Schiaparelli's network of fine lines, and those who claimed they were non-existent, or at best an optical illusion. The greatest proponent for the canals was Percival Lowell, a rich American who established the Lowell Observatory at Flagstaff, Arizona, especially to observe Mars. He did a great deal to further the idea (which Schiaparelli never intended) that the canals and the interconnecting dark round spots, which Lowell called Oases, were the achievements of intelligent life. His idea was that the canals had been built to irrigate the 'arid' reddish 'desert' areas of Mars by drawing water from the polar caps. He had many adherents to the idea, but his drawings bear no resemblance to the telescopic or photographic views of the planet.

In 1965, the U.S. Mariner 4 flight passed within 14,000 km (8700 miles) of the surface and transmitted back to Earth TV pictures and a wealth of data which were substantiated by the further Mariner 6 and 7 flights in the late 1960s. To everyone's great surprise the surface was found to resemble that of the Moon and was pock-marked with a great variety of crater-like formations. None of the photographs revealed the fine network of canals which had been drawn by Earth-bound observers. However, the so-called oasis-like features may indeed be representations of the largest craters. The illusionary telescope canals could then be explained as the eye tending to fill in detail between partially resolved craters at the limit of visibility.

The latest Mariner 9 flight revealed that it is now almost certain that fiery volcanoes once burnt over a wide area of the planet, and permanent dust storms may rage over certain regions. On the surface are huge valleys and wide cracks which indicate that strong erosional forces, including glaciation, have been at work and probably interior forces like those which cause tectonic plate movements on Earth.

Since the Mariner flights it has been realized that conditions for life on Mars are very extreme by earthly standards. The Martian polar caps are probably no more than thin layers of solidified carbon dioxide gas. At the poles the temperature must be *minus* 200 °C as compared with *minus* 130 °C in Antarctica. The atmosphere is much less dense than once thought and is equivalent to that at 30,500 m (100,000 ft) above the Earth. The Martian atmosphere is also rich in carbon dioxide and differs radically from the Earth's atmospheric composition and significantly contains much less nitrogen.

The darker regions, at one time thought by early observers to be seas, and then by later observers to be regions of primitive vegetation, are probably only discolorations due to variations of soil or rock. Differences in height of several miles do occur, but so far a detailed surface profile has not been resolved. Although craters are found over a wide area, the region round the feature called Hellas appears to be devoid of them. Significantly, few craters appear to have central peaks like those on the Moon, which may indicate differences in origin or in conditions prevailing at the time of origin.

Mars has two small satellites, Phobos and Deimos, which can only be observed in large telescopes. Phobos, the inner satellite, revolves in an orbit 9300 km (5800 miles) from the centre of Mars in a period 7 hrs 39 mins. This is a shorter interval than the planet's own revolution period which means that, as seen from Mars, Phobos rises in the west and sets in the east *twice* during the course of the Martian day. Observations made from the Mariner flights show Phobos to be an elongated object with a maximum diameter of 22 km (13·7 miles) and a low albedo (0·06), corresponding with the surfaces on the Moon and Mercury. Deimos is an even smaller object and probably not larger than about 6 km (3·7 miles). It revolves round Mars in a period of 30 hrs 21 mins at a radius of 23,500 km (14,600 miles). Both satellites were discovered during the opposition of 1877 by Asaph Hall with the 26-in refractor of the Washington Observatory. It is likely that both are captured asteroids (minor planets) which in the past approached Mars too closely and were perturbed into new orbits.

At the time of conjunction, when Mars is 380 million km (236 million miles) distant, its angular diameter is only 3″·5 and very little can be seen on its surface with even the largest Earth-bound telescopes. However, at opposition, when the angular diameter may be as much as 25″·1, small telescopes can reveal a great deal. At this time a magnification of ×75 will enlarge it to the apparent size of the full Moon seen with the naked eye. With a 2-in telescope some of the larger *maria* and either polar cap (depending on the season) can easily be observed. It is of interest that during the nineteenth century, R. A. Proctor made some remarkable drawings of Mars, using only a 1-in telescope and powers of × 50, × 100 and × 120.

The Asteroids and Bode's Law

In 1772, the German astronomer Bode (1747–1826) drew attention to a curious numerical relationship between the distances of the planets. Nowadays the relationship is called Bode's Law, but J. D. Titius of Wittenburg independently discovered the relationship at about the same time.

Take the numbers: 0, 3, 6, 12, 24, 48, 96, 192, 384, each of which (the second excepted) is double the preceding number. If the number 4 is then added to each, we obtain: 4, 7, 10, 16, 28, 52, 100, 196, 388, which in round numbers represents the approximate distance of the planets from the Sun expressed in radii of the Earth's orbit.

PLANETARY DISTANCES ACCORDING TO BODE'S LAW

Planet	Distance according to Bode's Law	True distance in astronomical units (A.U.)
Mercury	0·4	0·39
Venus	0·7	0·72
Earth	1·0	1·0
Mars	1·6	1·52
Asteroid Gap	2·8	2·9
Jupiter	5·2	5·20
Saturn	10·0	9·55
Uranus	19·6	19·2
Neptune	38·8	30·1
Pluto	77·2	39·5

Except for Mercury, Neptune and Pluto the agreement is remarkable. It will be noted that between Mars and Jupiter occurs a gap. It was this gap which led Bode to predict the discovery of a new planet and subsequently to attempt the organization of a band of observers to look for it. However, by a remarkable coincidence, on 1 January 1801, before the search was organized, Piazzi, at Palermo, noted a strange mag 8 object in Taurus. He observed it again on the following two nights, noting that it shifted position, and he continued to follow it for the next six weeks until an illness prevented him from making any further observations. Piazzi wrote to Bode, announcing that the body was a new tailless comet, but the object was now lost from view and considered unrecoverable. Nevertheless, by evolving a completely new computing method, J. K. F. Gauss (1777–1855), was able to predict its position to within $\frac{1}{2}°$ when it re-emerged one year later from the daytime sky, and he announced that the object was indeed the predicted planet with an orbit lying between Mars and Jupiter. At the insistence of Piazzi the new planet was called Ceres after the mythical fertility goddess of Sicily. Its mean distance is 2·8 A.U., but its diameter of 700 km (430 miles) was somewhat of a disappointment.

When looking for Ceres in March 1802, Olbers of Bremen discovered another new planet, which was later named Pallas. The search was intensified, and soon after, Juno and Vesta were added to the list. Astrophotography soon proved to be useful in minor-planet searches. By following the stars, a minor planet, because of its differential orbital motion, is recorded as a short trail and can easily be picked out later on the photographs. By such methods hundreds of minor planets have been discovered, and present-day estimates give a total population exceeding 50,000, not counting the very smallest objects.

Less than 2,000 orbits are known with any certainty, and most new objects are soon lost again. Many asteroids have highly eccentric paths which carry them near the Sun inside the orbit of the Earth. A number also have steep inclinations to the ecliptic.

Ceres is the largest asteroid, and by comparison with many of the others is almost a giant planet. Ceres and Pallas between them probably account for more than half the total mass, which does not exceed 1/3000 of the Earth's mass (*see* p. 77). The smallest objects probably represent the meteorites which collide with the Earth in numbers every day and are rock fragments a foot or so in size. More rarely the Earth collides with a larger asteroid, such as the body which fell some 25,000 years ago and gouged out the famous meteorite crater in Arizona, a kilometre (over 3000 ft) in diameter. Some other bright meteorites – whose orbits prior to collision have been accurately calculated – certainly originate from the asteroid belt, and the connection between the two is now well established.

Many asteroids, even when close to the Earth, are barely visible in the largest telescopes. When Icarus approached the Earth within 6·4 million km (4 million miles) in 1968, it showed up like a very fast moving faint star of mag 14. It was well observed, the new radar techniques being used, and the results give a rotation period of 2 hrs 16 mins, and a maximum diameter of 2 km (6600 ft). Icarus has a highly eccentric orbit which carries it round the Sun in a period of 409 days. At perihelion, 32 million km (20 million miles) distant, it roasts in a temperature over 1000 °C, while at aphelion, 283 million km (177 million miles) deep in space, its surface freezes at −250 °C. One of the closest minor planet approaches to Earth is made by Hermes, which in 1937 passed within 640,000 km (400,000 miles), but when the orbital geometry is suitable, it can approach closer than the Moon.

When Bode and Titius called attention to the gap between Mars and

Jupiter, it was announced that a sizeable body would be found. After the initial surprise on finding the small mass of Ceres, and when additional members were discovered, different theories were evolved to account for the situation. Olbers was the first to suggest that probably in the past a large planet had disintegrated owing to the close approach of another body. This theory is still known by the name 'Olbers' (hypothetical) Planet'. A variation of this theory reckons that initially there were twin planets which subsequently disintegrated because of self-destruction. Another idea suggested that asteroids were the left-overs from the early planet-building epoch. It is believed by many that at least the terrestrial planets were formed from the accretion of many smaller planets (planetesimals). Yet another theory links them with the comets.

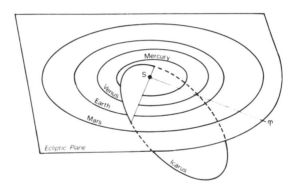

Fig. 21 The orbit of asteroid (minor planet) Icarus.

Observations show that some of the asteroids are irregular-shaped bodies, but variations in surface brightness could account for this. They certainly have no atmosphere, but they may accumulate surface ice mantles. Their gravitational pull must be negligible, and on the very tiny ones a man could literally jump off into space. Many of the smaller asteroids occur in family groups. There is one group called the Trojans whose orbit resonate with Jupiter. Near perihelion, if the Earth is suitably placed, many

THE ASTEROIDS

Asteroid	Number	Radius (km)	Mass* (g)	Mag	Rotational period (h)	(m)	Period (d)	a (A.U.)	e	i (°)
Ceres	1	350	60×10²²	6·0	9	05	1681	2·8	0·079	10·6
Pallas	2	230	18×10²²	6·0			1684	2·8	0·235	34·8
Juno	3	110	2×10²²	6·3	7	13	1594	2·7	0·256	13·0
Vesta	4	190	10×10²²	6·0	5	20	1325	2·4	0·088	7·1
Hebe	6	110	20×10²¹	6·6	7	17	1380	2·4	0·203	14·8
Iris	7	100	15×10²¹	6·7	7	07	1344	2·4	0·230	5·5
Hygiea	10	160	60×10²¹	6·4	18?		2042	3·2	0·099	3·8
Eunomia	15	140	40×10²¹	6·2	6	05	1569	2·6	0·185	11·8
Psyche	16	140	40×10²¹	6·8	4	18	1826	2·9	0·135	3·1
Nemausa	51	40	9×10²⁰	8·6			1330	2·4	0·065	9·9
Eros	433	7	5×10¹⁸	12·3	5	16	642	1·5	0·223	10·8
Davida	511	130	3×10²²	7·0			2072	3·2	0·177	15·7
Icarus	1566	1·0	5×10¹⁵	17·7	2?		408	1·1	0·827	23·0
Geographos	1620	1·5	5×10¹⁶	15·9			507	1·2	0·335	13·3
Apollo		0·5	2×10¹⁵	18			662	1·5	0·566	6·4
Adonis		0·15	5×10¹³	21			1008	2·0	0·779	1·5
Hermes		0·3	4×10¹⁴	19			535	1·3	0·475	4·7

* Earth = 5·98 × 10²⁷ g

of the larger asteroids can be seen in small telescopes and some even in binoculars and field glasses. Their current positions can be found by consulting suitable astronomical year books.

Jupiter

Outside the mass of the Sun, the only other significant mass in the solar system is Jupiter. It is by far the largest of all the planets, with an equatorial diameter of 142,800 km (88,700 miles). Its diameter measured through the poles is only 133,600 km (83,000 miles), and even the smallest telescope shows the polar regions to be significantly flattened. As a planet Jupiter is a very large body, and its mass exceeds that of the Earth 318 times. The mass of Jupiter has a profound gravitational influence on many bodies passing close by it, and the orbits of cometary bodies and asteroids are greatly affected.

Jupiter rotates on its axis faster than any other planet, taking 9 hrs 50 mins at its equator and 9 hrs 55 mins near the poles, which implies that the observed surface of Jupiter is not rigid. From our position on Earth, this outer surface, spinning at the equatorial rate of 40,250 km (25,000 miles) per hour, seems to be a very different kind of world from our own. Jupiter generally appears as a very bright mag −2·2 object and is brighter than Sirius. Its surface is an excellent reflector of sunlight – almost as good as Venus – which gives an indication of the kind of surface we are looking at. Through a telescope the most striking feature is the complex system of parallel belts of cloud that show up extremely well in photographs (Fig. 63). These so-called cloud belts develop irregularities and spots which are continually undergoing changes in both shape and colour.

The belts represent the outer region of Jupiter's extremely dense atmosphere. The most famous of all the atmospheric features is known as the Red Spot, although the name is a misnomer, for its colour is often anything but red. It appears as a huge oval shape (Fig. 63a), but unlike the other cloud-belt features, it maintains its shape over a long period of time. The Red Spot became a prominent object in the 1870s, although it had been observed as a less conspicuous feature long before this time. Many theories, some of them quite fanciful, have been put forward to explain the persistence of the Red Spot; nevertheless, no theory has yet proved entirely satisfactory. The great British amateur Peek, who observed the planet over many decades, considered that it was caused by a solid or quasi-solid body floating on top of the buoyant lower atmosphere.

Jupiter revolves round the Sun in a highly elliptical orbit with an eccentricity of 0·05. At perihelion it approaches the Sun to within 740 million km (460 million miles), while at aphelion it may be 966 million km

(600 million miles) away from the Earth. Its orbital velocity is 13 km per second, and it takes 11·86 years to make one complete revolution round the Sun.

Jupiter's mean density is only 1·3, an extremely low figure compared with the Earth (density 5·5), and is in fact lower than the Sun. Although the centre of the planet probably contains a dense solid core, the outer layer appears to be composed of various gases, including hydrogen in solid state under high pressure along with ammonia and methane. The surface temperature is about −100 °C, for Jupiter receives only a twenty-fifth of the Sun's radiation which the Earth receives. At the low temperature prevailing, the gas atoms and molecules composing the thick atmosphere cannot move energetically enough to escape the strong gravitational pull exerted by the planet.

One of the most puzzling features of Jupiter is its ability to emit radio waves in the decametre range. The origin of this radio emission is a great puzzle, but one of the principal contributory factors must be Jupiter's strong magnetic field and a radiation belt similar to the Earth's. Jupiter's magnetic field, however, is ten times larger than the Earth's and also (like the Earth) it is tilted to the geographical poles of rotation.* The magnetic field itself most likely owes its origin to the fast rotation period. One particularly puzzling feature is that the radio emission appears to be closely linked to the motion of the inner satellite Io. Radio bursts seldom occur except when Io crosses the plane of Jupiter's magnetic field on one side of the planet as seen from Earth. It has been noted that another kind of radiation also occurs near the poles of Jupiter which does not appear to be influenced in any way by the movement of the satellite. This is termed the 'A' source radiation. The Io radio-wave component is termed the 'B' source. Although at the present time we do not know the fundamental cause of these emissions, they are definitely connected with Jupiter's magnetosphere, which is much like the Earth's, but on a far grander scale.

The size of Jupiter has led to speculation about the behaviour of the planetary matter immediately below the atmosphere. Jupiter appears to be at the critical theoretical maximum size for a planetary body. If it were larger, it could begin to radiate as a star-like object. If it were denser, the inner core would become degenerate matter. Jupiter in fact may be a unique example of a cosmic body half-way between a planet and a star.

Even the smallest telescope shows Jupiter's system of cloud belts. At opposition, when the disc has an apparent diameter of 50″, a magnification of only ×40 will show it as large as the full Moon appears with the naked Jupiter's tilt is 8°.

eye. Also easily seen are the four Galilean satellites, so named because of their discovery by Galileo, in 1610, although Simon Marius is reputed to have seen them nine days earlier. They are plainly visible with theatre glasses, like small bright pearls strung out in line, supposing that one or more of them is not passing behind the planet or in transit across the surface. All four are comparatively large bodies as satellites go. Ganymede and Callisto are both larger than the Moon, and Europa and Io are comparable in size. Jupiter also has eight other satellites, but these are much smaller in diameter. The twelfth satellite was discovered in 1951 and is probably not larger than about 30 km (19 miles) and only visible in the largest telescopes. The outer four satellites are of great interest, since they move in retrograde, or clockwise, orbits which is the opposite direction to the usual planetary and satellite motion. The view from Jupiter of the system of 12 moons circling round, some in a contrary direction, must be intriguing, although the outer ones will shine as relatively faint objects even from Jupiter's surface.

It is reasonable to suppose that Jupiter has many other satellites so tiny that they are well beyond visibility from the Earth. Perhaps all but the principal Galilean satellites are captured asteroids; from a dynamical point of view it would be easy for Jupiter, owing to its large very influential mass, to sweep up a whole family of these tiny bodies during the course of its long history. The Galilean satellites are much better reflectors of light than our own Moon, indicating that their surfaces may contain areas of frozen gases. Observations with large telescopes show some dusky and bright features which appear to be semi-permanent. The smaller asteroidal satellites are less dense and may be composed of part rock and a significant part solidified gases – or they may be small rocky nuclei (cores) with very thick ice mantles.

The large satellites can be observed in transit across the disc of Jupiter particularly by the dark shadows they cast on the surface. A 3-in telescope shows a surprising wealth of detail both of the cloud belts and of the satellite transits. They can also, of course, be eclipsed by Jupiter, and variations in the eclipse times led Roemer, in 1675, to arrive at the first accurate determination of the speed of light (*see* p. 36).

Saturn

As the second largest of the Jovian planets, Saturn holds a premier position owing to its spectacular rings, which even in modest-sized telescopes are revealed as extensions to either side of the disc at their widest opening. When Galileo first looked at Saturn, he thought he saw three bodies, for his primitive telescope was not able to resolve the actual shape of the ring feature, and it was left to Huygens, in 1656, with his long-focus aerial telescope, to announce its true nature.

Saturn is even more oblate than Jupiter. The equatorial diameter is 119,400 km (74,150 miles) and the polar 106,900 km (66,400 miles), a direct consequence of a rapid rotation period of 10 hrs 14 mins. Its mean distance from the Sun is 1,427,700,000 km (887,000,000 miles), and in dealing with such a large linear quantity it is better expressed as 9·5 A.U. (9·5 times the Earth–Sun distance). Because of the eccentricity of its orbit (0·056) it varies in distance 160 million km (100 million miles) between perihelion and aphelion. As seen from Earth, during its orbital revolution of 29·59 years its apparent diameter changes from 14″ to 20″.

Apart from the uniqueness of its ring feature, Saturn is similar to Jupiter in many ways. However, its density of 0·7 is lower – even less than water. Any solid matter within its distinctly oblate sphere must be very near the centre. Hydrogen appears to be a very dominant constituent and makes up between 60 and 70 per cent of the bulk, plus appreciable quantities of methane slightly in excess of Jupiter, but containing less ammonia. Owing to the low surface temperature of −143 °C, it has been concluded that much of the ammonia is frozen out of its atmosphere. The atmospheric depth appears to be unusually thick, and internal pressure must approach 50 million (Earth) atmospheres.

Saturn's surface cloud belts are less distinctive than Jupiter's. The equatorial region seems to be very white and brilliant by contrast to the darker yellowish subtropical and temperate regions, while the polar regions often appear quite greenish. The equatorial regions occasionally develop distinctive white-spot features which often last for many weeks, such as the spot which was seen in 1933 (Fig. 142). However, the normal lifetimes for irregularities on Saturn's cloud belts are measured only in days or at best a few weeks.

The rings of Saturn have always been a great puzzle and a constant source of wonder. They never fail to attract attention and are among the rare astronomical phenomena that, from a lay-public point of view, live up to their reputation when seen through a telescope! The rings lie exactly in the plane of Saturn's equator, which is inclined at 28° to the Earth's orbit. This plane, like all the planetary axial planes, is fixed in space so that from the Earth we see a gradually changing appearance of the rings as Saturn revolves round the Sun in its 29·5-year period (Fig. 22). At the position when the Earth passes through the exact line of sight of the ring plane, they disappear from view for a short period (only a day or two with the largest telescopes). This indicates that they must be extremely thin (approximately 20 km [12 miles]) and is compatible with the idea that the rings are composed of individual fragments of matter revolving round Saturn as miniature satellites in gravitationally stable orbits. A 'solid' monolithic ring or halo of material is not physically possible.

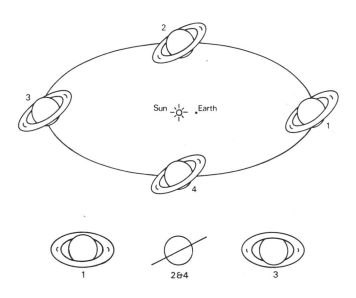

Fig. 22 The changing appearance of Saturn's rings as seen from Earth. Position near 3 corresponds to the appearance in Fig. 64a. Positions 2 and 4 refer to the appearance when the Earth passes through the exact line of sight of the ring plane.

The 'solid' appearance of the ring is an effect created by the close proximity of the tiny particles (once believed to be made of frozen ammonia [NH_3]) viewed at the great distance from the Earth. They are in some instances partially transparent to starlight, particularly near the edges, as would be expected of such a composite arrangement. But the structure of the rings, as a whole, is not uniform. There are three distinct divisions or zones, known as A, B and C. Between A and B occurs a gap known as the Cassini's division (Fig. 64a), so named after G. D. Cassini (1625–1712), one of its first observers. This division is plainly visible with as little as a 2-in telescope × 25 when the rings are at their widest opening, directed towards the Earth. The inner ring, known as the 'crepe ring', is more readily seen by the shadow it casts across the surface of Saturn.

The particles forming the ring must revolve round the planet at variable speeds depending on their distances out. The nearer particles must accomplish their revolution in about 5 hrs 50 mins, and the most distant in about 12 hrs 5 mins. The rings must be constantly in a state of flux, and over a period of time undergo observable evolutionary changes. Their origin is still unsolved. One popular idea is that they once formed a satellite which approached Saturn too closely. Under tidal distortion the satellite broke up into a myriad of particles that now constitute the ring and which continually suffer further fragmentations owing to frequent collisions with each other as they jostle for position. It has been pointed out that a similar event could take place with our own Moon. Nevertheless, others consider this idea too glib, and reason that the ring material represents the 'left-overs' which did not accrete to form Saturn during the planet-building stage. Present estimates indicate that the ring material does not exceed 1/25,000 of the bulk of Saturn, and more positive information about their nature awaits the deep-space grand-tour planetary probes planned for the late 1970s.

Saturn has ten known satellites. The largest, Titan, exceeds the size of Mercury and it appears to possess a considerable atmosphere much like the parent planet. The smallest satellite, Janus, was discovered in 1966 when, in December that year, the ring system was seen 'edge on' to the Earth.* Janus lies about 150,000 km (93,000 miles) out from the surface of Saturn and is probably not larger than about 300 km (200 miles). The outermost satellite, Phoebe, has, like Jupiter's outer satellites, a retrograde orbit, but it is a much larger body than those of Jupiter and most likely represents a large captured asteroid. As with Jupiter, Saturn also probably possesses many tiny satellites, unobservable from the distance of the Earth owing to their extreme faintness, and may, like the ring system, be composed largely of frozen ammonia or other frozen gases.

Uranus

Uranus is rarely visible to the naked eye and was therefore unknown to the astronomers of the ancient world. It was discovered by the musician, amateur-astronomer William Herschel on 13 March 1781 with a

* During the present decade the rings are closing until 1980.

self-constructed 7-in diameter reflecting telescope. At first it was thought to be a new comet, but its motion across the sky soon indicated that it was a completely new planet lying beyond the orbit of Saturn, and which unsuspectingly had been observed on numerous previous occasions by telescopic observers who had unwittingly catalogued it as a star.

Its mean distance is 19·2 A.U., and it revolves round the Sun in a period of 84·02 years, during which time its distance varies between 2590 and 3150 million km (1608 and 1956 million miles). Uranus is about 3·7 times bigger than the Earth, its equatorial diameter equals 47,600 km (29,660 miles), and its polar diameter about 2860 km (1780 miles) less, indicating that the degree of polar flattening is equal to that of Jupiter's. The rotation period is 10 hrs 49 mins, but quite unlike all other planets its axis of rotation lies almost in its orbital plane inclined at an angle of 98°, thus implying a retrograde motion. Theoretical astronomers have long puzzled over this peculiar situation, but no satisfactory solution has yet been suggested.

At its nearest approach, Uranus has an apparent diameter of $3''\cdot1$ so that in a large, or even a moderate (6 to 12-in) telescope the planet can be seen as a greenish disc. Markings (if any) are very indistinct, showing that the planet is covered with a thick blanket of atmosphere. The physical constitution of Uranus closely resembles that of Jupiter and Saturn, although it has a slightly higher density (1·60). The surface regions are primarily hydrogen plus methane which is about ten times more abundant than in Jupiter. The surface temperature cannot be less than -160 °C, which accounts for the spectroscopically observed feature that gaseous ammonia is absent from the atmosphere and that methane is extremely abundant.

Although faint parallel cloud belts, or bands, similar to those of Jupiter, have been reported from time to time, the French visual astronomers at Pic du Midi in the Pyrenees, who have excellent planetary observing conditions, consider that such reports are unfounded. Uranus has five satellites, which revolve in steeply inclined retrograde orbital paths in the same plane as the planet's revolution axis. The largest, Titania, is almost 1000 km (625 miles) in diameter.

Uranus can easily be picked out in the night sky using theatre glasses, or better still, 8 × 30 prismatic binoculars. If one uses a star atlas (e.g. *Atlas Coeli*) showing stars to mag 7·7, Uranus, which shines at mag 5·8–6·5, can then be spotted as a stranger against the background stars. But its daily position must first be ascertained from the *Astronomical Ephemeris*, an annual publication usually obtained in the reference section of a public library.

84

Neptune

Neptune is well beyond naked-eye visibility and was discovered after a systematic search was made for it, subsequent to discrepancies noted in the observed position of Uranus. Two mathematical astronomers, Le Verrier in France, and John Couch Adams in England, independently calculated the position of the hypothetical body, and both came up with the same solution. Observational discovery was made by Galle at the Berlin Observatory in 1846, on the same night as receiving Le Verrier's ephemeris, when a mag 8 star-like object was almost instantly recognized as the new planet. This proved a great triumph for mathematical astronomy, and the full, detailed story makes absorbing reading. Soon after discovery, it was found that many observers had previously seen Neptune (like Uranus) but unknowingly catalogued it as a fixed star.

Neptune lies beyond the orbit of Uranus at a mean distance from the Sun of 30·06 A.U. and has an orbital period of 164·8 years, which means that it has not yet made a complete revolution round the Sun since the time of discovery. In many ways it is a sister planet to Uranus, for it physically resembles it. However, Neptune is slightly larger, with an equatorial diameter of 48,400 km (30,000 miles) and a polar diameter of about 900 km (560 miles) less.

The surface of Neptune has a high albedo of 0·84 owing to a thick atmosphere of hydrogen and appreciable quantities of methane gas. As would be expected of a planet with significant polar flattening, it has a fast rotation period of 15 hrs 30 mins. The axis of the planet is inclined at 29° to the equator, but its orbital inclination to the ecliptic is only 1°·8.

Telescopically the disc of the planet has only an apparent diameter a little over 1″ as seen from the Earth, and not much can be noted on its bluish coloured surface. It may be located in the sky with 10 × 50 binoculars or smaller telescopes by plotting its ephemeris position* on a star chart depicting stars to a limiting magnitude of 9·0 (e.g. *Atlas Eclipticalis*, the B.D. or *Webb*).

Neptune has two satellites. The fainter one, Nereid, discovered in 1949, has the most eccentric of all satellite orbits in the solar system, and its distance varies 1,335,758 km to 9,817,000 km (830,505 miles to 6,096,368 miles) from Neptune. Triton, the brighter satellite, discovered in 1846, a few weeks after the planet, is also remarkable, for it possesses a retrograde orbit inclined 40° to Neptune's orbit and 20° to the planet's equator, but in contrast to Nereid, its orbit is almost circular.

* Gleaned via reference libraries from the *Astronomical Ephemeris*, *BAA Handbook* or *Sky and Telescope*, etc.

Pluto

After irregularities in the motion of Neptune had been noted which could not be accounted for by Uranus, it was proposed that yet another planetary body lay further out in space far beyond the orbit of Neptune.

During the first quarter of the nineteenth century calculations were made by various astronomers to predict the position of the new object, as had been done and which led to the later discovery of Neptune. Such a calculation was made by Percival Lowell and Pickering. Lowell died in 1916, but the observers at Flagstaff Observatory, which Lowell founded, had begun a systematic search in 1905 which continued intermittently until 1931 when Clyde Tombaugh, a one-time farmboy amateur astronomer who had recently turned professional, discovered a new object in the constellation of Gemini. It was found in a spot within a few degrees of that predicted by Lowell, and apparently it looked like a further triumph for celestial mechanics. However, after a period of observation, it was realized that Pluto (it was commemorated by using the initials of Percival Lowell) was only a small body and probably not larger than 4000–6000 miles in diameter. Unless this small body had an unbelievably high density it could not possibly produce detectable perturbations on Neptune. General opinion now has it that the discovery of Pluto was a remarkable coincidence rather than a brilliant feat of mathematical astronomy.

Another theory, however, conjectures that Pluto *is* a larger body than the Earth, but owing to what is termed specular reflection* its measured (apparent) diameter is much smaller than its true diameter.

Pluto revolves round the Sun in an orbit taking 248 years. But its orbit is markedly different from other planets since it has an unusually high eccentricity (0·247). At aphelion it lies at a distance of 39·5 A.U., while at perihelion it can be as little as 29·7 A.U., which is less than Neptune's distance from the Sun. For several decades since 1969, Pluto will in fact be closer to the Earth than Neptune (Fig. 45). However, there is no chance of an orbital collision between Pluto and Neptune, for the path of Pluto is inclined at a relatively steep 17° angle.

Photometric observations reveal that Pluto varies in brightness in a regular pattern, and from this a rotation period of 6 days 9 hrs 16 mins 54 secs has been derived.

One theory accounts for the origin of Pluto by considering that it is an ex-satellite of Neptune which somehow broke free and moved along in an independent orbit. Another idea considers it to be the brightest member of a whole family of smaller planets lying at the far boundary of the solar

* When only light from the centre of a sphere is reflected.

system, and which are too faint even to be photographed. Pluto itself shines at mag 14·5 and requires at least a 15- to 18-in telescope in order to glimpse it visually.

COMETS AND METEORS

Comets

Comets are curious nebulous bodies which, like the planets, revolve round the Sun, but generally have more eccentric orbits. The name comet is derived from the Greek *Kometes,* meaning 'hairy'. In ancient times, only the brighter comets visible to the naked eye could be seen. These comets generally develop long, spectacular tails when they near the Sun, and the Greeks likened them to flowing female tresses.

In the ancient and medieval western world, the sudden appearance of a bright comet caused much consternation among the populace, for they had considerable astrological importance. All manner of evil influences were ascribed to them: they were presages or omens and signified the impending or recent death of a king or pope, or alternatively they were signs of pestilence, plague and war. Superstition dies hard – if it dies at all – for when Halley's comet last returned to the Sun in 1910, patent-medicine vendors in the U.S.A. did a brisk trade in country districts selling comet pills to ward off the evil influence of the hairy star. In 1970, during the Egyptian–Israeli conflict, the appearance of Bennet's comet was received with apprehension by the *fellahin,* who first thought it was sent by the Almighty as a weapon to aid the Israelis.

In early times Aristotle had pronounced that comets belonged to the regions of the atmosphere and were therefore *not* truly astronomical bodies. This view was held by most until 1577 when Tycho Brahe proved that the comet of that year lay at a distance far beyond the Moon, and so they officially entered the domain of astronomy.

For accurate description and observation of ancient comets, we have to rely heavily on Chinese records, for all the observations originating from the western world are coloured by astrological overtones. Even well-educated men described the appearance of comets in the most hysterical unscientific manner. A famous example is that provided by the Paris physician André Paré, who in describing the comet of 1528 says:

'. . . This comet was so horrible, so frightful, and it produced such great terror in the vulgar, that some died of fear and others fell sick . . . it was the colour of blood . . . on both sides of the rays of this comet were seen a great number of axes, knives, blood-coloured swords, among which were a great number of hideous human faces, with beards and bristling hair.'

It was also the Chinese in A.D. 837, whose observations are contained in the annals of Thong, who first noted the significant fact that comet tails always point *away* from the direction of the Sun. This fact was not commented upon in the West until Peter Apian independently noted it in 1531.

Nowadays the vast majority of comets observed are not visible to the naked eye. Spectacular comets are rare examples, and the average comet appears like a small faint spherical cloud, usually without a tail. Indeed, many comets come and go which are too faint to be seen visually even in the very largest telescopes, and they are only recorded by lengthy photographic exposures.

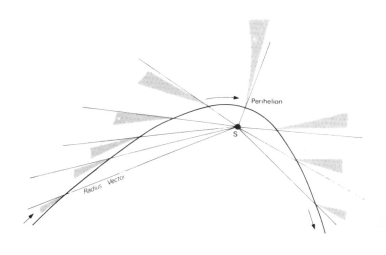

Fig. 23 Comet tails, especially the spectacular 'dust' tails, always point away from the Sun, but they generally lag a little behind the Comet–Sun line (or radius vector).

The average bright comet consists of three main components: the nebulous *coma*, with or without central condensation; the *nucleus*; and the *tail*. The coma is usually round but it may also be elliptical or directionally distorted in some way. Generally speaking, the coma is brighter towards the centre and this is referred to as central condensation. The central nucleus is a more luminous region and is often star-like in form, and which no amount of magnification will expand into a planetary-like disc. Dimension-wise the diameter of a comet's head may range in size from a thousand kilometres to several hundred thousand kilometres. The tail may consist of either (or sometimes both) a 'dust' tail or a gas tail. The spectacular comets are the ones showing impressive 'dust' tails which curve away like graceful broad plumes. The gas tails are less bright and are usually straight and very narrow, but they are sometimes observed as highly contorted features (Fig. 65).

Cometary orbits round the Sun have very elongated shapes (Fig. 24), which in some instances stretch out into deep space, almost half-way towards the nearest stars. Comets are divided into two main classes of object chiefly by the size of the orbit, but this leads to physical differences as a result of their orbital evolution. Although it appears likely that *all* comets originated from a common physical parent stock, they now have marked variations.

The short-period comets are a class of comet which have been perturbed from previous long-period orbits and captured by Jupiter when at some time in the past they made a close approach. Jupiter's bulk is very significant in perturbing comets, whose mass and densities are extremely small by comparison to planets, which led an American astronomer to coin the description of a comet as: 'a great big bagful of nothing'. From a long-period orbit measured in some thousands of years, a comet may be transformed into one with a period of only 8–12 years. Once under the dominance of Jupiter, the comet may undergo further, very rapid, changes in orbit at each revolution, but these evolutionary changes can be predicted with great accuracy. The majority of the short-period comets are faint objects and only rarely become naked-eye objects.

The long-period comets have orbital periods measured in some hundreds or even thousands of years. The demarcation separating short- and long-period comets is somewhat arbitrary. For example, among the short-period comets some include Halley's comet, which has a period of about 75 or 76 years. The most significant feature of long-period comets is their greater brightness and certain differences in physical make-up. But many long-period comets are not bright objects as seen from the Earth since their orbital geometry often does not allow them to approach the Sun close enough. It is reasonably certain that some comets are never seen at all, for they come to perihelion at the vast distance of Jupiter's orbit. Occasionally, however, we do catch a glimpse of one, and Humason's comet (Fig. 65) was such an example, which underwent some strange variation in brightness and tail structure.

Among the long-period comets is a group known as the 'Sungrazers'. This highly descriptive name is applied because these comets literally appear to graze the outer atmosphere of the Sun at perihelion passage. They are prevented from dashing headlong into the Sun owing to the significance of Jupiter's mass and the combined masses of the other planets influencing the position of the gravitational centre of the solar system. This (barycentric) point lies a little outside the Sun, but its position varies depending on the disposition of the planets as they revolve in orbit.

Most 'Sungrazers' are large, extremely brilliant comets which can often be seen in broad daylight. Comet Ikeya-Seki 1965 VII, discovered by the two Japanese amateur astronomers whose joint names it now bears, was seen at midday very close to the Sun's surface at the time of perihelion passage, shining at mag −9. When it later passed into the dawn sky, it developed a tail over 40° long. However, tails may be much longer, and at one time the tail of the Great Comet of 1843 stretched across the entire heavens.* These great tails consist of 'dust' particles and may be formed owing to the radiation pressure of sunlight on the nucleus or quasi-nucleus in the cometary head. The gas tails are composed of ionized gases (plasmas) caused by the interaction of the solar wind on the comet's head. For this reason comets have been nicknamed 'space barometers', for they are able to register the strength of the solar wind by its effect on the length, brightness and structural behaviour of the gas tail.

All comets – like planets – obey the fundamental laws of gravity, and likewise their orbits have the properties of a conic section, i.e., circular, elliptical, parabolic or hyperbolic (Fig. 24). However, the description of cometary orbits can be confusing. Comets are often spoken about as being parabolic (open-ended orbit) which would imply that such objects are nonperiodic bodies, for they could not possibly return to the Sun again. The anomaly is brought about by the fact that when a comet is discovered, and its orbit computed, it is easier to calculate an orbit in terms of a parabola even though we know it is probably an ellipse. In the case of a long-period comet, its orbital arc described near the Sun is almost identical whether one assumes a parabolic path or an elliptical one. Only when a large number of observations have accumulated over a longer length of the orbital arc, is it possible to calculate a definitive orbit. In practice the least likely orbit that

* In terms of true length: equal to the distance of the orbit of Mars.

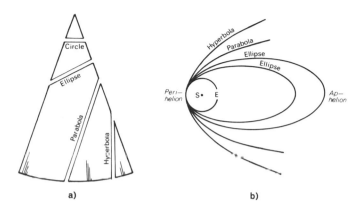

Fig. 24 Conic sections and cometary orbits:

(a) A cone can be sectioned to produce a circle, an ellipse, a parabola and hyperbola.
(b) Near perihelion the different cometary orbits are often indistinguishable.

a comet may possess is actually a parabola, for it would be quickly perturbed into either an elliptical or a hyperbolic one.

All the comets that have ever been observed have belonged to the solar
system; therefore all must have elliptical orbits. Nevertheless, some
comets have been observed whose definitive orbits have shown that they
were travelling in hyperbolic paths. This would imply that the comet did
not belong to the solar system and had arrived here from some other
stellar system. At one time indeed this view was held, but it was later found
that *all* hyperbolic orbits are induced from inside the solar system through
dynamical interaction between the comet and a planet. These comets can
never return to the Sun again and are therefore lost to the solar system for
all time.

Halley's comet is one of the best known objects in the entire sky. It
bears the name of Edmund Halley (1656–1742) who first discovered, as a
consequence of Newton's gravitation laws, that comets revolve in periodic
orbits. By checking the past apparitions of comets, he formed the opinion
that the comets seen in 1531, 1607 and 1682 were actually the same comet,
and he predicted that it would return again in 1758. The prediction was
fulfilled, and it was seen again in 1835 and in 1910, and it is due to return in
1985–6. In more modern times the appearance of Halley's comet has been

traced in various records back to 240 B.C. One of the most famous returns in early history was in 1066, when it was recorded on the Bayeux tapestry (Fig. 66). Owing to the influence of Jupiter, the comet does not have an exact orbital period but may be retarded a year or more between each apparition.

Although the orbital motion of comets is well understood, there is little conclusive evidence available about their physical nature or their origin. One idea suggests that a comet is a mixture of ice-conglomerate consisting of a highly porous mass of solidified gases including water-ice, ammonia, methane and possibly carbon dioxide and dicyanogen plus solid particles. This mixture supposedly forms a discrete nucleus which, triggered by solar radiation sublimates to form the gaseous coma. This model has been appropriately nicknamed 'the dirty snowball'. An alternative idea considers that the cometary particle clouds are formed when the Sun occasionally passes through a cloud of cosmic dust and gas during the course of its 225-million-year revolution round the Milky Way. By various dynamical processes these particles are impressed to form smaller compact clouds, and they remain as semi-permanent members of the solar system. This model is nicknamed 'the flying sandbank'. The chief difference between the two ideas is that the 'sandbank' idea does not envisage a discrete nucleus but rather a pseudo-nucleus suggested by the greater concentration of cloud particles towards the gravitational centre of mass.

Chemically the two basic models are not very different – only the *modus operandi* of their formation. To account for comet origins 'the dirty snowball' concept utilizes the idea of an exploding Olbers' (hypothetical) planet (p. 76) which supposedly, at the same time, formed the asteroids. In this theory, the comets represent the concentration of lighter elements which were perturbed out to the far boundaries of the solar system where the fragments remain until they are perturbed inwards again and appear near the Sun as comets. A third theory considers that comets represent volcanic material ejected from the Jovian planets, or their satellites, but on dynamical grounds this idea seems unlikely, and it has few adherents outside the Soviet Union.

Comets are certainly very puzzling bodies, and their chemistry is not easy to interpret directly, although we know they are rich in hydrocarbons and carbon monoxide. When a bright comet nears the Sun, the spectroscope reveals lines of neutral iron, nickel, chromium, silicon, manganese and calcium, plus ionized lines of calcium. Also very prominent are the yellow sodium lines which may give the comet a yellowish colour. Observations of comets are restricted by the limitations of the Earth's atmosphere. It was therefore very interesting when the orbiting astronomical observatory

(OAO 2), launched on 7 December 1968, first observed comet Tago-Sato-Kosaka in January 1970. It made the surprising discovery that the comet was enveloped in a vast hydrogen cloud about a million miles in diameter. This feature can only be seen in ultraviolet light which is normally absorbed by the Earth's atmosphere. The discovery of this vast hydrogen cloud surrounding a comet gives a foretaste of what may be accomplished when a space probe is dispatched to observe the head of a comet at close quarters. Such programmes are envisaged and will likely occur during the next decade, for it is doubtful if any definitive physical knowledge of comets can be gained from further observation through the Earth's atmosphere.

Comet Discovery

When a comet is found, it is identified by the name, or names, of its discoverer(s) and by an alphabetical letter following strict sequence in any particular year – *a*, *b*, *c*, etc., so that the comet designated Comet Humason 1961*e*, means that the comet was discovered by the astronomer Humason and was the *fifth* comet discovered in the year 1961. At a later date, usually during the following two years, when full orbital details are known, the comet is renumbered in the chronological order that it came to perihelion in that year, or in a later year, and affixed with a Roman numeral, i.e., I, II, III, etc. Thus comet Humason is entered in the permanent catalogue: Comet 1962 VIII Humason 1961*e*, as the eighth comet in 1962 to come to perihelion (which one immediately sees it did not do in the year of discovery). If a comet is independently discovered by a number of people, it is designated by the names of the first three observers to report it, e.g., Tomita-Gerber-Honda 1964*c* (VI). If the comet discovered is unusually bright, as may often happen, and seen simultaneously by many observers, no names are designated, and the comet is known by its year and Roman number or, even more frequently, simply by the expression 'Great', thus 'The Great Comet of 1843' (1843 I [a]).

Each year the number of comets under observation varies between about 10 and 20 (*see* table p. 94), but in 1951 no fewer than 22 comets were under observation at the same time. Periodic comets can be recognized in the catalogue nomenclature by the prefix P/, thus Comet 1962 VII P/Tempel (2) 1961*b* was the thirteenth apparition of Tempel's second periodic comet, first discovered in 1873 (1873 II).

Comet hunting is a useful field of pursuit for amateur participation. Many new comets are regularly discovered by amateur observers who specialize in such activity and use short-focus wide-angle optical equipment such as 25 × 105 binoculars (Fig. 136) or *f*/4 'richest field' telescopes.

Professional discoveries in the modern era are usually accidental ones, made when photographing star fields for other research purposes.

NUMBERS OF COMETS

Year	New Comets Discovered	Periodic Comets Recovered	Total Comets under Observation
1964	3	5	17
1965	5	5	19
1966	4	2	9
1967*	4	10	20
1968	7	4	16
1969	5	5	19

* A record year for the combined total of comets discovered and periodic comets recovered.
Note : Since many comets have orbits which allow them to be observed over periods often exceeding a year, it follows that the total number of comets under observation during any particular year will usually exceed the total number of discoveries and recoveries of that year.

EXAMPLES OF SHORT- AND LONG-PERIOD COMETS

Comet	No. of Appear-ances	Period (Years)	Peri-helion Distance (in A.U.)	Eccen-tricity	Orbital Inclina-tion
1970e P/Encke	48	3·30	0·338	0·847	12°·4
1969e P/Honda-Mrkos-Pajdusakova	4	5·21	0·556	0·815	13°·2
1967d P/Tempel (2)	14	5·27	1·391	0·543	12°·4
1964i P/Holmes (lost for 60 years, found again 1964 and 1970)	5	7·35	2·347	0·379	19°·5
1909c (1910 II) P/Halley	29	76	0·587	0·967	162°·2
1907b P/Grigg-Mellish	2	164	0·923	0·969	109°·8
1965f Ikeya-Seki*	1	880	0·008	0·999	141°·8
1882d Daylight Comet*	1	1000(?)	0·008	0·999	142°·0
1843I Brilliant Comet*	1	1000(?)	0·006	0·999	144°·3
1961e Humason	1	2900	2·133	0·989	153°·2
1956h Arend-Roland†	1	10000(?)	0·316	1·000	119°·9

* Sungrazing comets. † Slightly hyperbolic orbit.

ROMAN NUMERAL DESIGNATIONS OF COMETS IN 1969

Comet	Date of Perihelion	Name	Discovery Year and Letter
1969 I	Jan. 12.2	Thomas	1968j
II	Apr. 19.0	P/Gunn	1970p
III	May 10.8	P/Harrington-Abell	1968i
IV	Sept. 11.0	P/Churyumov-Gerasimenko	1969h
V	Sept. 23.0	P/Honda-Mrkos-Pajdusakova	1969e
VI	Oct. 7.6	P/Faye	1969a
VII	Oct. 12.4	Fujikawa	1969d
VIII	Oct. 29.1	P/Comas Sola	1968g
IX	Dec. 21.3	Tago-Sato-Kosaka	1969g

Comets may be found in any part of the sky, but the spectacular ones are often found close to the Sun near the time of their perihelion passage. Chances, therefore, of discovering a bright comet are then greater if a search is carried out of the western sky shortly after darkness and the eastern sky shortly before dawn near the ecliptic zone. Chances are much improved in subtropical and tropical climates where the twilight period is minimal, and the sky above the horizon is much darker immediately after sunset and shortly before dawn. It must be borne in mind that although comets near perihelion are often moving at velocities of over 300 km (190 miles) per second, this is not directly apparent to the observer. A comet can be observed to move in relation to the backcloth of stars only after an interval of time. There is much confusion by the lay public in believing that comets, like meteors, are swift-moving objects whose move ment is immediately *apparent* to the naked eye.

Meteors

The common name given to a meteor is 'shooting star'. This is, of course, simply a convenient description of the visible appearance of a meteor when it is seen in the night sky. They are not stars but tiny fragments of cosmic matter which enter the Earth's atmosphere in countless millions every day and because of their high velocity burn up owing to sudden frictional braking.

Nowadays there are a number of classes of meteor, and this term is used only loosely to describe a range of cosmic fragments. The visible 'shooting star' is called a *meteoroid* and is not to be confused with a *meteorite* (p. 98). Most meteoroids are vaporized in the upper atmosphere in less than a second (Fig. 93). What we observe is actually the trail of ionized gases left as the tiny fragment has passed by. Generally these trails only last for a fraction of a second, but some persist for much longer periods. Meteoroid particles average approximately the size of a grain of sand or less. If we count the number visible with the naked eye on a clear moonless night, about six will be seen every hour. If we observe through wide-angle binoculars, such as 10 × 80s, in any given field of view, we also observe about the same number of fainter telescopic meteors caused by particles smaller than those which produce the naked-eye variety. Fewer telescopic meteors are seen through normal telescopes, since their angular field of view is very limited.

Meteoroids are thought to be genetically related to comets, for in some instances they have identical orbits in space. These orbits intersect the Earth's own orbit and each year, as the Earth passes through, it gives rise to an annual display of 'shooting stars' or a 'meteor shower'. Such a display is that of the Leonid meteors – so named because on 17–18 November each year they appear to radiate from a point inside the 'sickle' of the constellation of Leo (the Lion [see Sky Map, p. 187]). The Leonid shower is connected with the orbit of comet Tempel (1866 I), which has a period of 33·18 years. Although the shower is an annual event, the major displays, often termed 'meteor storms', occur at intervals of 33 years. There was a major display in 1799, 1833, 1866, 1899, and a poor one in 1933. After 1899 it was considered that the main stream of meteoroid particles had been perturbed away from the Earth's orbit, but in 1966 there occurred the greatest display of all time. As the eastern hemisphere was wrongly positioned at the exact time of this event, only those observers in the western United States and eastern Siberia were treated to this memorable spectacle, for at one period the rate of entry was estimated to exceed 300,000 meteoroids per hour, and the sky was literally ablaze with 'shooting stars'.

There are other annual displays which can be observed (Table below), but the meteoroid rates are fairly low, as can be seen in the column for expected rates per hour. Some of the annual showers do not have any known observable comets associated with them. The Geminids which have an orbital period of only 1·65 years, approach the Sun to within 0·14 A.U.

THE MAJOR ANNUAL METEOR SHOWERS

Shower	Date of Peak Activity	Radiant Coordinates RA	Dec	Duration of Detectable Meteors	Duration of Peak Days	Expected Hourly Rates
Quadrantids	Jan. 3	231°	+50°	1–4 Jan.	0·5	50
Corona Australids	Mar. 16	245	−48	14–18 Mar.	5	5
Virginids	Mar. 20	190	00	5 Mar.–2 Apr.	20	5
Lyrids	Apr. 21	272	+32	19–24 Apr.	2	10
Eta Aquarids	May 4	336	00	21 Apr.–12 May	10	20
Ophiuchids	June 20	260	−20	17–26 June	10	20
Capricornids	July 25	315	−15	10 July–5 Aug.	20	20
Southern Delta Aquarids	July 29	339	−17	21 July–15 Aug.	15	20
Northern Delta Aquarids	July 29	339	00	15 July–18 Aug.	20	10
Pisces Australids	July 30	340	−30	15 July–20 Aug.	20	20
Perseids	Aug. 12	46	+58	25 July–17 Aug.	5	50
Kappa Cygnids	Aug. 20	290	+55	18–22 Aug.	3	5
Orionids	Oct. 21	95	+15	18–26 Oct.	5	20
Southern Taurids	Nov. 1	52	+14	15 Sept.–15 Dec.	45	5
Northern Taurids	Nov. 1	54	+21	15 Oct.–1 Dec.	30	5
Leonids	Nov. 17	152	+22	14–20 Nov.	4	varies
Phoenicids	Dec. 5	15	−55	5 Dec.	0·5	50
Geminids	Dec. 13	113	+32	7–15 Dec.	6	50
Ursids	Dec. 22	217	+80	17–24 Dec.	2	5

They also appear to consist of extraordinarily dense particles which tend to be decelerated far less in the Earth's atmosphere – much less than other meteoroids. No comet has been discovered with such a short period, and this indicates that it may now either be extinct, or no concentration of meteoroid material even took place to form a cometary component. The structure of a meteor stream *cannot* be likened to a continuous 'bicycle tube' of material revolving in orbit. The mass and concentration of material is very variable. Particles are spaced at extremely wide distances. Even at the peak of the previously mentioned Leonid displays each meteoroid must have been separated from its nearest neighbour by at least 15 km (9 miles). Although circumstances point towards the idea that meteoroids represent cometary debris that is spread out in a kind of orbital wake, the evidence is not conclusive. In certain meteor streams which give rise to annual showers, the associated comet *follows* the meteoroid particles. Nevertheless, it is beyond dispute that comets and meteoroids associate in the same orbit, but how their roles are interwoven is not yet fully understood.

When radar techniques were developed during World War II, it was found that the ionized trail left by a meteoroid could be 'observed' by bouncing off a radar beam. This development led to the discovery of a number of daylight showers hitherto unobservable by naked eye or by wide-field photography. One of the daylight showers may be associated with Halley's comet (Eta Aquarids).

Micrometeorites (*and Micrometeoroids*) represent much smaller cosmic particles than the meteoroids and are only a few microns* in diameter. They enter the Earth's atmosphere and fall to the Earth, arriving at the surface intact. Whatever cosmic velocity they initially possessed in outer space is reduced considerably before they reach the ablation zone of the lower atmosphere, through which they can then pass creating little or no frictional heat. The very smallest particles take a considerable time to reach the Earth's surface, and many simply drift about in the thermal upcurrents for an indefinite period. There is some evidence to suppose that such particles form the nuclei for the formation of noctilucent clouds (Fig. 40).

Dust collected from the upper atmosphere by high flying aircraft or rockets has also contained tiny black spherical particles which probably have a terrestrial origin and are the product of industrial processes connected with steel and iron smelting. The true micrometeorite has a loose, fluffy, irregular consistency and is very friable. They generally lack crystal structure, which is the result either of cosmic radiation damage or perhaps lack of sufficient binding energy in low gravitational fields as is likely to exist during their formative stage. The term *micrometeorite* is normally applied to particles completely unaffected by their passage through the atmosphere, while *micrometeoroids* are those particles which have partially melted owing to some frictional heat being created during their fall.

* One micron = 10^{-3} mm.

These particles are probably the true representatives of cometary decay processes rather than the larger meteoroids and may be the shorn-off fragments from the cometary nucleus, or from the larger meteoroid particles eroded by the effects of the solar wind. On rare occasions, when the Earth passes through the orbital plane of a comet, a very narrow stream of such particles can be observed. Sometimes these observations have given rise to speculation about so-called anomalous comet tails which point *towards* the Sun. However, this is simply an illusory effect of geometrical perspective. The phenomenon was even noted by the keen-eyed Chinese in ancient times. It was observed in the comet 1851, and in the spectacular Arend-Roland comet 1957*b* (Fig. 145).

Meteorites and Fireballs

Meteorites are solid cosmic bodies, which, like meteoroids, have entered the Earth's atmosphere from outside, and until recent times they were the only source of cosmic material available for detailed study in terrestrial laboratories. Unlike meteoroids they possess sufficient size to carry them down to the Earth's surface, and only part of their mass is vaporized or dissipated during their passage through the atmosphere.

In flight they take on the form of brilliant fireballs and have a dramatic appearance; they are also known as *bolides*, from the Greek meaning 'to throw'. However, not all fireballs become meteorites, for the definition, meteorite, implies that some part of the object reaches the Earth's surface, and many brilliant fireballs of insufficient initial mass appear to burn themselves out while still at considerable height. In some definitions, a bolide is the name given to a fireball which is seen to explode in the sky. The criterion which distinguishes a meteoroid from an ordinary fireball is brightness. In terms of stellar magnitude (p. 110) a fireball is classified as such when its brightness exceeds mag -4 (approximately the brightness of Venus).

Both meteorites and fireballs may produce events visible in full sunlight and leave dust trails in the atmosphere which may persist for several hours. It is likely that a number of large meteorites hit the Earth's surface each day, but the majority go unnoticed, for they fall in remote areas and in the oceans. When a major fall occurs in an inhabited area, the phenomenon observed is very impressive. If it occurs at night, the fireballs may have sufficient brilliance (for a fleeting instant) to light up the sky as if it were broad daylight. After the visible effect is over, loud rumbling noises follow which resemble the firing of cannons, thunderclaps and the now familiar everyday noise caused when an aircraft breaks the sound barrier. The interval of time elapsing between the visible and acoustic effects depends on the distance away from the observer at which the event occurs. Frequent other strange phenomena accompany the passage of the fireballs, such a noise which has been variously described as a 'hissing', 'rustling' and 'the singing of wind through telegraph wires'. These noises are often noted *ahead* of the other acoustic effects and probably arise owing to an electro-magnetic or electrophonic cause, for they travel at great velocity. Animals appear to be particularly receptive to their frequency, and numerous instances are quoted when horses, cattle and dogs have suddenly become frightened or agitated a few seconds *before* a local meteorite event occurred.

But although many meteorites fall to the Earth's surface every day, a fall in a particular locality is a rare astronomical event. In modern times there are no substantiated records of death caused to a human being, although there are well-cited instances of minor injuries and also damage to property. It has been calculated that in a country such as the U.S.A. (pop. 200,000,000) a human being will be struck by a meteorite once every 9000 years, so it can be seen that the chances are extremely remote.

The most famous example of a large meteorite crater resulting from a fall is the Barringer crater located in the Arizona Desert, near Flagstaff. The hole measures three-quarters of a mile in diameter and is 180 m (600 ft) deep. Estimates give a probable date of occurrence some 25,000 years ago. Scattered over the Earth's surface are even bigger, but less documented, examples, such as the Chubb crater in Ungana, Canada, about 5 km (3 miles) across, and considerably larger structures whose possible origins are still debated by geologists.

Falls providing fragments less than 100 kg are reported in detail quite regularly in the scientific press. In Britain a spectacular fall occurred on Christmas Eve 1965, at Barwell, a village in Leicestershire, from which a total of 50 kg of material was recovered. The recovery of this material, scattered over a wide area in the village, created great interest, for the British Museum instituted a scale of monetary rewards for material handed in. The difficulty involved in the recovery of this valuable collectors' material led in later years to a Bill being presented in the British Parliament to allow for all meteorites falling in Britain to be subject to the laws of treasure trove such as with gold and silver finds. Small bodies such as the Barwell meteorite often fragment while still in the atmosphere and scatter the material over a wide area which is often elliptical in shape. After such a fall, pieces are found lying on the surface, and only the heavier chunks, weighing a kilogram or more, become buried in the ground at a depth depending on the penetrability of the ground on which they fall. With a small meteorite the braking effect of the atmosphere is such that the meteorite is slowed down at a height of about 12 km (7 miles) so that it loses the ability to create sufficient frictional heat to continue glowing; it subsequently falls to the ground under the influence of gravity with all its initial cosmic velocity completely lost.

Meteorites can be classified into three main classes of object.

(a) The irons, which are magnetic and contain quantities of nickel.

(b) The stones, made up of rocks rich in silicates.

(c) Stony-irons, a transitional mixture of stones and irons.

All the principal classes can be subdivided into various minor classifications.

It is not easy to make positive identifications of meteorites by appearance alone, especially if they have lain exposed to the weather for any length of time. Only by using several criteria is it possible to be sure that it is not simply a terrestrial rock or some other artefact. Many so-called iron meteorites are specimens gleaned from blast furnaces or are pyrite nodules picked up in exposed chalk country such as Salisbury Plain in England (Fig. 70). Usually a genuine meteorite has a distinctive black fusion crust, the result of frictional melting during its passage through the atmosphere. In rare instances the crust is transparent like glass. If the meteorite has suffered considerable fragmentation, internal parts, of course, will show no crust.

The black fusion crust also contains highly distinctive thumblike marks called regmaglypts, again the result of frictional surface melting during its passage through the atmosphere. They are able to provide indication of the original size of the meteorite. Each thumblike depression is reckoned as being one-tenth the dimension of the parent body. Iron meteorites, when analysed, give very positive indications of their identity. The nickel content of such meteorites ranges from 5 to 20 per cent. whereas terrestrial iron contains nickel in less or greater quantities. If the smoothed surfaces of iron meteorites are etched by acid, they will reveal a curious Widmanstatten pattern (Fig. 72) which cannot be simulated in terrestrial laboratories. This pattern is the result of how the crystalline structure has formed owing to slow cooling in space over a long period of time.

Meteorites, and perhaps all fireballs, seem to be associated with the asteroid belt, for their orbits are very similar. When the Pribram meteorite fell in Czechoslovakia in 1959 (brightness mag -19), and the Lost City meteorite in the U.S.A. in 1970, their orbital trajectories were photographed in precise detail, and provided sufficient information to show them as typical asteroidal objects with paths overlapping that of the Earth. Both were also shown to be very old and at least comparable to the age of the Earth.

Various techniques have been developed for estimating meteorite ages. One method is through the slow decay of radioactive isotopes whose half-lives are accurately known. These estimates provide the chronological age from the time when the meteorite actually formed in space. Cosmic rays can also be utilized to indicate how long a particle's outer surface has been exposed to bombardment, which will often provide a different result from the former method, showing that fragmentation has taken place in space between formation and the time immediately before the object entered the Earth's atmosphere.

One of the most interesting sub-varieties of stony meteorites are the carbonaceous chondrites. These contain significant amounts of hydrocarbon material which has given rise to speculation about whether organic life exists elsewhere in the solar system. Claims have been made at various times that organic material has been positively identified in such rare meteorites, but the position is complicated owing to the difficulty of preventing such objects from suffering contamination after they fall to the surface of the Earth. The problem is by no means confined to meteorite samples, as all lunar rock samples suffer similar contamination in spite of the ultra-clinical precautions which are taken beforehand.

Tektites

There is a special kind of object called tektites whose origins have not yet been decided with complete certainty. They are found in specific geographical locations throughout the world (Fig. 25) and have been recognized as separate objects for two centuries. They consist of small glassy objects (70–80 per cent SiO_2) which at first glance superficially resemble the terrestrial volcanic rock obsidian. However, on closer inspection, they are found to be quite different and have the appearance of small distorted globules of liquid rock which has solidified during flight.

In fact the name tektite is derived from the Greek *tektos* meaning 'molten'. No terrestrial rock *in situ* has yet been found which resembles them, and neither do they bear any affinity to the local country rocks in the various locations where they are found lying on the surface.

They have a range of very diverse shapes such as tear-drops, rods, discs, flanged buttons or even dumb-bells. However, all these shapes can be explained as a result of aerodynamic modelling as they were solidified during flight. Their colours range through black, brown to a kind of bottle-green, and in size they range from a walnut (or smaller) to an apple.

Their origins could be explained by two basic processes:

(*a*) That they are terrestrial in origin and may be a result of volcanic activity.
(*b*) That they are a product of meteorite impact—either in the form of modified cosmic material or crustal material transformed by heat.

Weight of opinion at present favours the meteorite origin. The discovery of microtektites (diameter 1 mm) in the South Pacific marine sediments, dated at approximately 700,000 years, which resemble the nearby Australian mainland tektites, lends support to this idea.

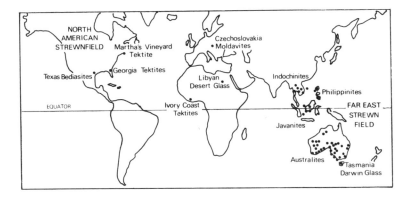

Fig. 25 Geographical distribution of tektites.

Zodiacal Light and Gegenschein

As the name suggests, the zodiacal light occurs in the same part of the night sky as the Zodiac, i.e. the region immediately extending either side of the ecliptic. It takes the form of a very diffuse, elliptical-shaped, soft pearly glow which is widest towards the direction of the Sun and gradually narrowing until it merges into the background illumination of the night sky. Most favourable times for observation (in the northern hemisphere) are on clear, moonless nights shortly after dark in the spring, or before dawn during the autumn. The advantage of observing at these periods is that the ecliptic is more steeply inclined, projecting the cone of light higher in the sky and on to a darker background (Fig. 26). In temperate zones of either hemisphere the zodiacal light is difficult to observe in summer owing to prolonged twilight skies. In tropical or equatorial regions the ecliptic always presents a steep angle to the horizon, and the sky darkens more rapidly after sunset; as a consequence the zodiacal light can be seen throughout the year. In desert climates, with transparent air, it becomes a conspicuous feature in the sky, and is often referred to as 'the false dawn' since its appearance suggests the sky at the onset of dawn. The Australian Aborigines have long known it by the name 'pickaninny daylight' – little daylight – and legends about it are woven into their folklore.

The *gegenschein*, or counterglow, occurs in the night sky diametrically opposite the position of the Sun. It is much less easy to see, and this can be illustrated by the fact that its existence was not recognized until the middle 1850s, after which it was discovered independently on at least four subsequent occasions. It takes the form of an elliptical patch or cloud measuring about 10° by 20° and with its broadest width lying along the ecliptic. In temperate latitudes it is an elusive phenomenon to spot, but by using averted vision, it certainly can be picked out when the sky transparency is at its best, e.g. lack of moonlight, and unusual atmospheric clarity such as occurs after a rainstorm.

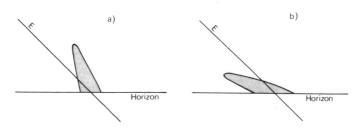

Fig. 26 The appearance of the zodiacal light cone in the western evening sky in northern temperate latitudes:
(*a*) In spring.
(*b*) In autumn.

Spectroscopic examination of the zodiacal light shows it to be caused by the reflection of sunlight from small solid or quasi-solid particles of matter concentrated along the ecliptic plane of the planetary orbits. The origin of the particles is uncertain. They may represent debris associated with meteoroids and comets, or, they may be unformed left-overs remaining after planetary accretion. The lifetime of this material is important in the understanding of its origins. The Poynting-Robertson effect considers that small particles in space absorb heat via solar radiation, and the re-emission of energy creates a retarding force which is proportional to the velocity of the particle; as a result a small particle will eventually lose orbital velocity and spiral into the Sun. The time-scale of this occurrence is about 100,000 years, which is a very short period in relation to cosmological time. If then the

Poynting-Robertson effect is a real one, it means the zodiacal material must be constantly replenished. However, the Poynting-Robertson effect may only be significant on truly spherical particles of matter, and we know from recovery of micrometeorites that a substantial fraction are very irregular structures. Another little understood, but very influential factor might be the solar wind. At sunspot maximum there could be a 'pushing away' or 'spiralling out' of material further into space, while at sunspot minimum, when the solar wind drops to a 'breeze', there would be a spiralling inwards. Thus the zodiacal particles could be shuttled back and forth, and they could remain in space for much longer periods than the presently accepted understanding of the Poynting-Robertson effect allows for.

The theories accounting for the gegenschein are even more conjectural. The most likely explanation, however, is that it represents a cloud of 'dust' (similar to the zodiacal 'dust') situated at a gravitational libration point.

Such libration points, or Lagragian points, named after the French mathematician-cum-astronomer Lagrange (1736–1813), have also been investigated in the dynamics of the Earth-Moon system. There is good reason to suppose that the combined gravitational pull of Earth and Moon can trap clouds of interplanetary dust. Theoretically, this could occur at any of five libration points, but in practice it probably only occurs at three. Such clouds were reported from specially conceived aeroplane flights in 1966–8 from an altitude of 12 km. The observers reported visual sighting of small clouds 2°–4° in size at the limit of naked-eye visibility and fainter than the gegenschein. Photographic evidence has also been produced, but this is far from conclusive. At the present time the existence of such clouds must remain problematical, and their further investigation awaits the extraterrestrial photoelectric space-probe scans of the next decade.

Space Probes and Orbiting Extra-terrestrial Observatories

Since the initial space probes of the 1960s, knowledge of the nearer members of the solar system has increased far beyond that previously obtained by Earth-bound telescopic observation. It may be said in fact that during the last decade, more information has been gleaned about the terrestrial planets than was previously known from over 300 years of dedicated optical study. The limitations imposed by the Earth's atmosphere make it unlikely that any further significant contribution to planetary science can be made from the surface of the Earth.

In the future, much routine work will be accomplished by orbiting astronomical observatories (OAO* Fig. 77) and non-orbiting astronomical

* Also: OGO (Orbiting Geophysical Observatory); OSO (Orbiting Solar Observatory); MOL (Manned Orbital Laboratory).

observatories established on the lunar surface. Their use in conjunction with electronic image intensifiers (p. 239) will be principally in stellar research. Such extraterrestrial observatories should then be in range of mag 30 objects as compared with the present Earth-bound Mt Palomar 200-in telescope limit of mag 24.

The investigation of objects lying within the solar system is suitable for space probes using traditional rocket propulsion fuels, or by newly developed nuclear-reaction propulsion which can be tapped in either electric or thermal energy forms.

Various missions beyond Mars, and also for non-planetary objectives, have long been planned in great detail. Of particular importance, but with only low *military* priority, is the proposed mission to a comet. A space probe through the head of an active comet offers the best chance of deciphering the nature of these enigmatic objects. First step is a fly-by within 1000 km (600 miles) of the head with a prime objective of carrying out a detailed spectroscopic examination and direct television scans of the physical structure. The next step will be the attempted landing on a comet nucleus, if such a monolithic structural feature is actually found to exist. Among the known periodic comets only three offer favourable opportunity at their returns between the present time and the middle 1980s. This includes the long-awaited scheduled return of Halley's comet in 1986.

At the end of the 1970s, and at the beginning of the 1980s, will occur a unique planetary configuration suitable for exploitation by a deep-space probe. During this period the outer Jovian planets will be lined up so that a probe could visit in turn both Jupiter and Saturn; or at a schedule timed a little later, include Jupiter, Uranus and Neptune. On this second opportunity an auxiliary probe could be launched during flight which would fly-by Pluto.

Such far-ranging probes depend much on the future uncertain space plans of the U.S.A. and the U.S.S.R. However, nearer to home, the establishment of lunar bases would appear to be closer to practical realization. Such problems that exist concerning the establishment of lunar bases are mainly long-term biological ones. If manned stations are developed, the major biological problem will be one of the maintenance, and support, of life systems in a low gravitational field. We already know the disastrous effects to man that short-term exposure to weightlessness may incur. At present the long-term effect on human life in the Moon's low gravitational field is not yet known. Purely technical problems, such as the extraction of oxygen, water and subsequently hydrogen from the lunar soil, can be said to have been theoretically solved. Such operations may become as routine a chore on a lunar base as melting ice for domestic purposes on polar expeditions. All heat and electrical energy could be produced through the direct tapping of incident solar radiation. Even the chemical refuelling of rocket batteries on the lunar surface would not seem to present any serious difficulties.

That observatories outside the Earth are necessary for future astronomical research is irrefutable. Nevertheless, at the present time the cost of launching a *single* OAO into Earth orbit would provide three 200-in Palomar-type observatories. At the moment, owing to limited space budgets, rapid advance in this field is unlikely, for tactical spin-off from the OAOs has only limited military application.

Recent Space Probes

The U.S. spacecraft Pioneer 10 successfully reached Jupiter early in December 1973, passing within a distance of 130,000 km. Colour pictures and important new data on Jupiter's magnetic field were relayed back to Earth. Pioneer 11 is scheduled to reach Saturn in October 1979. After the great success of the Pioneer 10 mission, Pioneer 11 was re-targeted so that after passing Jupiter it will pass close to Saturn to probe the nature of the planet's ring system (*see* Fig. 76).

The U.S. Mariner 10 spacecraft made a successful fly-by of Venus on 5 February 1974, passing within 5760 km, when it relayed back to Earth very detailed information about the atmosphere of the planet. Hydrogen was found to be a very important element in controlling the chemical balance. Magnetometer readings indicated a weak magnetic field approximately 1 per cent of that of the Earth.

Following its rendezvous with Venus, Mariner 10 flew past Mercury's dark side at a height of 700 km on 29 March 1974. The most surprising discovery was the detection of a magnetic field. Spacecraft TV cameras confirmed the belief that Mercury's surface is cratered and similar to that of the Moon; it has been suggested that the surface features are primordial and date back some 4 to 4·5 aeons.

Soviet space probes Mars 4 to 7 arrived at the planet between 10 February and 9 March 1974. Although new data was obtained, equipment failures led to these missions falling short of expectations. The Soviet probes confirmed that Mars has a magnetic field several hundred times weaker than that of the Earth.

III THE STARS AND THE MILKY WAY

The Stars

The fact that the stars are separated from the planets by an immense distance was quite unknown in the ancient world. Only after a long period of time was it slowly realized that the inverted bowl of the sky was not simply a flat surface, without depth, but a great open vault extending away in space to limits whose distance was immeasurable.

The planets could be seen to move over the backcloth of stars and, like comets, were noticed to show appreciable parallax in relation to the stars when viewed from separate locations on the Earth's surface. The stars, however, remained fixed and apparently immutable, and the only sign of change was the rare appearance of a bright nova which would then slowly fade away until lost again from view.

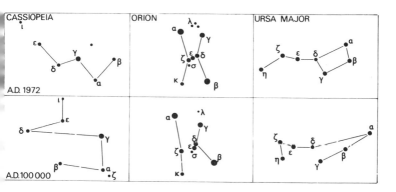

Fig. 27 The appearance of well-known star groups today and 100,000 years hence.

In 1718, Edmund Halley showed for the first time that some of the brighter so-called fixed stars like Sirius, Procyon and Arcturus actually moved in relation to neighbouring stars. But this shift was owing to proper motion, and although it strongly hinted that the brighter stars were most probably the nearer ones, the discovery gave no clue to the actual distance at which they lay. The independent proper motion of stars is very small, and during the course of many centuries the appearance of any particular constellation remains the same. Only in intervals measured in several thousands of years do the familiar constellation patterns appreciably alter (Fig. 27).

After Halley, many observers pursued the problem of finding stellar distances. Even Bradley, the third Astronomer Royal, whose observations detected the phenomenon called *nutation*, failed – as did Sir William Herschel with his large reflecting telescopes (Fig. 78). All these observers correctly surmised that if observations of a particular star field were made at intervals of six months (thus using the diameter of the Earth's orbit round the Sun as a trigonometrical base line), a displacement of the brighter stars against the fainter (background) stars should be apparent (Fig. 28). But although Herschel failed as Bradley had done, his observations, like Bradley's, led to another significant discovery. Some of the stars that Herschel had observed were double stars, and he found that over a period of time, some of the pairs appeared to revolve round one another in periods estimated at several years.

It was not until 1838 that the first parallax was successfully detected and measured almost simultaneously by three observers using three different stars. Although the accuracy that they achieved was inferior to modern determinations, it was the breakthrough which had been long awaited. A yardstick had been provided by which the nearer visible Universe could now be measured.

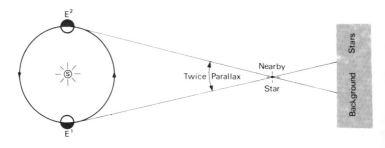

Fig. 28 Stellar parallax. The trigonometrical parallax of a nearby star is determined by using the diameter of the Earth's orbit as a base line. A star is observed from two diametrically opposite positions spaced six months apart, i.e. E¹–E².

PARALLAX DETERMINATIONS

Star	Parallax	Dist. (A.U.)	Modern Obs. Parallax	Dist. (A.U.)
α Centauri	1·8	200,000	0·750	270,000
61 Cygni*	0·314	640,000	0·285	700,000
α Lyrae	0·262	760,000	0·100	2,000,000

* The apparent shift of 61 Cygni is equal to 1/5000 the apparent diameter of the Moon.

The Night Sky

Even a casual glance at the night sky reveals that the stars are not evenly distributed about the heavens. The brightest stars form distinctive shapes and patterns, which the astronomers of the ancient world divided into groups or constellations to which they attached mythological names and legends (Fig. 80). The date of the actual naming of the constellations is lost in history, and it is not certain which civilization began the custom. We know, however, that the Babylonians used much the same constellations as we do today (Fig. 82), and they were passed on through the centuries by the Egyptian (Fig. 83), Greek and later Arab astronomers (Fig. 151). The Chinese and Polynesians independently devised their own constellations, but these were not known to the western world until modern times.

Nowadays there are 88 recognized constellations distributed throughout both hemispheres (see Sky Maps and p. 244). In addition to the traditional star names passed on chiefly from Arab sources, the stars are designated by Greek or Roman letters, or by special letters or numbers referring to certain astronomical catalogues (see p. 246, star nomenclature). Thus the Pole Star, or *Polaris*, which is the brightest star in the constellation of Ursa Minor (Latin for Little Bear) is known as α Ursa Minoris.* Most of the brighter naked-eye stars, and those of special interest, are designated in the method described; the fainter stars and those beyond naked-eye visibility can be identified by their celestial coordinates RA and Dec (p. 34).

Stellar Distances

In all linear measurement some kind of unit length has to be decided upon which is suitable for the distances involved. Distances on the Earth's surface are measured in miles or kilometres. In the solar system distances can be measured in similar units, but when they reach large numbers, long strings of noughts become almost meaningless. They can, of course, be expressed in powers of ten, e.g. the distance of Mars from the Sun equals 227,940,000

* The genitive form is used in this instance.

kilometres or 227·9 × 10⁶ km. A more convenient unit is expressed in terms of the astronomical unit (A.U.), the mean distance between the Earth and the Sun, or the *radius vector* distance of the Earth's orbit. Using the example of the distance of Mars again, this would be 1·52369 A.U.

However, for purposes of measuring stellar distances other units need to be adopted; for even one of the nearest stars previously quoted, 61 Cygni, lies 64,800,000,000,000 miles distance from the Earth. The speed of light provides a convenient unit for their distances. Light travels at 300,000 km (186,000 miles) per second, and this distance is accordingly known as a *light second*. In one year it travels 9,500,000,000,000 km (6,000,000,000,000 miles), which equals one *light year*; thus the distance of the star 61 Cygni can be expressed more conveniently as 10·8 l.y. Although the unit light year (l.y.) is still used a great deal, in professional astronomy distances are related directly to parallax. A star, for example, which gives a parallax of 1″ (one second of arc) is denoted as being at a distance of one *parsec*, and one parsec is equal to 3·258 light years. But even the parsec is sometimes too small a unit, particularly for very distant objects. Therefore, for very large distances we use the *kiloparsec* (a thousand parsecs) and also the *megaparsec* (a million parsecs).

Only the nearer stars can be measured by the parallax method. For distances greater than about 200 l.y. (70 parsecs) other methods have to be adopted. One such way is by a measure of the star's brightness. If the star's real brightness, or luminosity, can be deduced, then by measuring its apparent brightness, its distance can be calculated. It is known from physical laws that the apparent brightness of any light source is inversely proportional to the square of its distance from the observer. The true brightness of a star, in turn, can be deduced from analysing the physical properties of the star (p. 112). The *absolute magnitude* of a star denotes the brightness or magnitude the star would appear if placed at a distance of 10 parsecs, or in other words corresponding to a parallax of 0″·1. By adopting a common yardstick of 10 parsecs and ignoring their true distances, individual star luminosities can be directly compared, and this is particularly useful in statistical studies of star population.

Stellar Magnitude

Observation of the night sky will reveal that the stars are of unequal brightness. The brightness, or magnitude, of a star depends on its distance away and its luminosity. It was the Greek astronomer Hipparchus who first conceived the idea to divide all the visible stars into six arbitrary grades of brightness. The brightest group, about twenty in number, were called first magnitude stars. Ptolemy later refined the method and introduced fractions of a magnitude.

This magnitude system was satisfactory until the invention of the telescope. But by the middle of the nineteenth century, as telescopes penetrated further into space, and exact comparisons between stars were needed, a more scientific method was required. After much discussion, the magnitude system was related to a precise logarithmic scale. Instead of guesswork each magnitude differed from the next one by a factor of 2·512; thus a first magnitude star was 2·512 times brighter than a star of second magnitude. The difference then between five magnitudes corresponds *exactly* to a ratio 100:1. Other alterations were also necessary: some of Hipparchus' first magnitude stars needed to be uprated so that the concept of zero (0) and minus (−) magnitudes was introduced. Thus a star of magnitude 0 (zero) is 2·512 times *brighter* than a first magnitude star, and a −1 magnitude star is 2·512 times *brighter* than a 0 magnitude star. The brightest star in the sky, Sirius (α Canis Majoris), is mag −1·4. About 3000 stars are visible to the naked eye on any particular clear, moonless night. With 8 × 30 binoculars, mag 9 stars are just visible and number about 200,000. The faintest stars photographed with the 200-in Mt Palomar telescope are mag 24·0, and in numbers uncountable. In the future, by the use of electronic intensifiers (p. 239), magnitudes of 30·0 may soon be brought within reach. On the scale of stellar magnitudes the Sun's brightness is reckoned as −26·7, and the Moon's −12. Care must always be taken to distinguish between a star's *apparent* magnitude (as we see it at its true distance from Earth) and its *absolute* magnitude (as the star would appear at a distance of 10 parsecs).

The Nature of Stars

Owing to the great distances of even the nearer stars, we are not able to observe them as resolvable discs like we do with most of the planets. The largest telescopes show them only as point sources, but by the use of the spectroscope in combination with other methods, a great deal of information can be gleaned about their physical make-up. Additionally, we can study the Sun close at hand and compare its characteristics with the distant stars, for nowadays there is good reason to suppose that the Sun is an average, typical example of the kind of star that makes up most of the population of our own galaxy, the Milky Way.

Nevertheless, among the total population of stars is represented an enormous variety both in size and in temperature (Fig. 84). Some stars are extremely massive objects with diameters 3000 times larger than the Sun. Other stars are dwarfs which emit so little radiation we can barely see them at all. Surface temperatures vary from less than 1500 °C to over 500,000 °C. Some of the giant stars are cool spheres of gas with tenuous atmospheres while other stars, particularly many of the dwarfs, are dense bodies which give rise to phenomena not yet fully understood by present-day physical laws.

Although the stars are extremely diverse, they conveniently fall into definite arrangements denoting size, brightness and temperature in a kind of periodic table. One star arrangement is called the main sequence, and our own Sun fits conveniently in at a precise slot. But besides the main sequence stars are the various other groups of stars known as White Dwarfs, Subgiants, Giants and Supergiants. All can be shown in a definite relationship if they are plotted on a special diagram known as the Hertzsprung-Russell (H-R) diagram, so called after the Danish and American astronomers who conceived the basis for its construction (*see* Fig. 87). Every star can be plotted on the diagram by its absolute magnitude along the ordinate axis, and its spectral class (or colour) along the abscissa.

The H-R diagram is an extremely useful tool in the study of stellar evolution. All stars, like all other objects in the cosmos, undergo evolutionary changes during the course of their life-spans, which are all reflected on the H-R diagram. The fact, however, that the majority of stars appear to lie in the main sequence and red giant regions, shows that they probably spend a greater proportion of their evolutionary time there.

Physics of Stars

Like the Sun the stars are spheres of gas and produce their energy through internal thermonuclear reactions of various kinds. The kind of nuclear reaction depends on what stage they are at in their evolution. The most important of the reactions are the carbon-nitrogen cycle and the proton-proton reaction. All the elements contained within the Universe have been evolved in the stars via various linked nuclear reactions which begin with the basic material hydrogen that forms a substantial proportion of a star's bulk.

The actual physical condition, or state, of a star can be deduced through the examination of its spectrum in a similar way that we do with the Sun. Most star spectra are indeed similar to the solar spectrum with its background of bright continuum* crossed by the complex of dark absorption Fraunhofer lines. Some stars have bright-line emission spectra as well, but other stars have only bright-line spectra and no dark absorption lines. There are stars rich in hydrogen and helium, others, like the Sun, rich in

* Red, orange, yellow, green, blue, indigo, violet.

[*continued p. 20*

Fig. 29 Egyptian sky goddess Nu (or Nut).

Fig. 30 Avebury Circle, England, as depicted by William Stukeley about 1725.

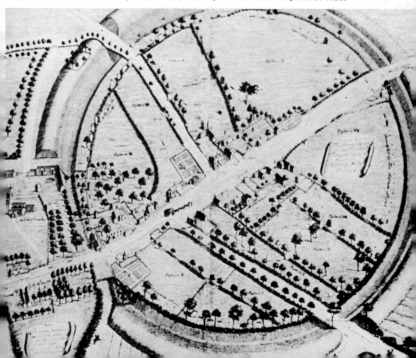

Fig. 31 Stonehenge, England. Remains of a megalithic observatory.

Fig. 32 Egyptian 'star clock' from the Tomb of Rameses VII.

Fig. 33 Babylonian cuneiform tablet containing astronomical tables.

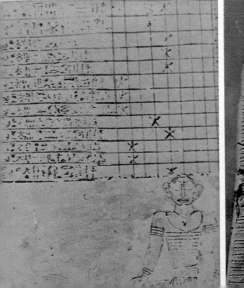

Numbers indicate hours of day a–12 noon Greenwich time (Zero Meridian)

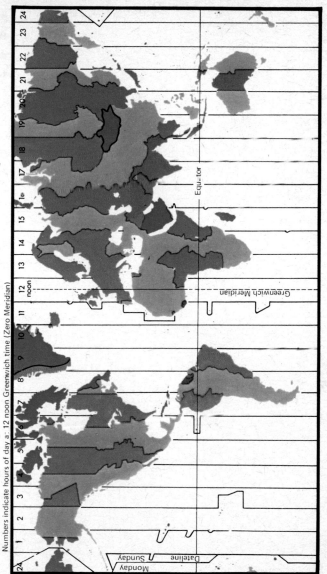

Fig. 34 Time zones of the world.

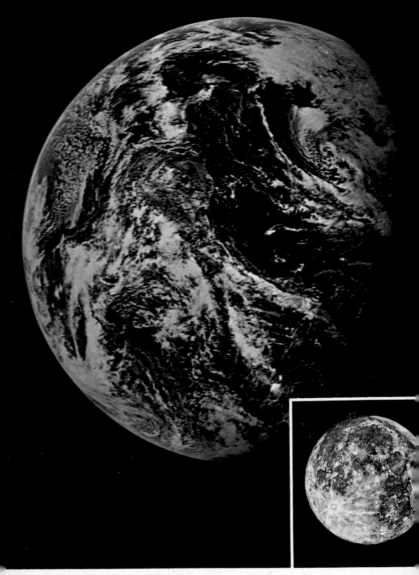

Fig. 35 Relative sizes of Earth and Moon.

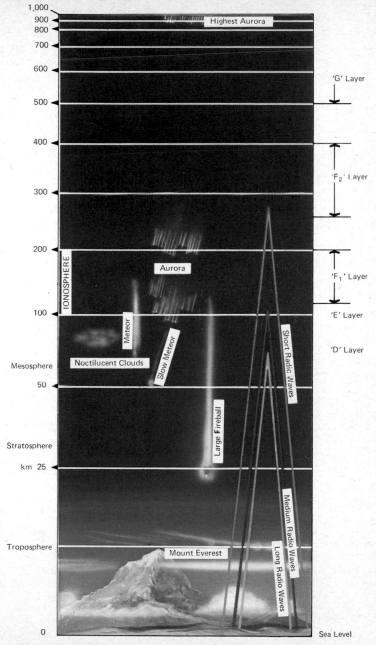

Fig. 36 Section through the Earth's atmosphere.

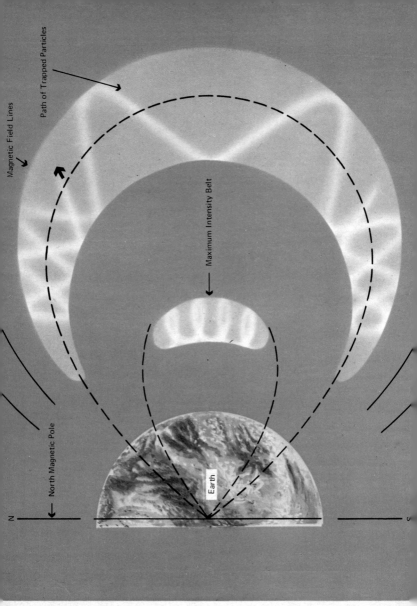

Fig. 37 The van Allen Belts.

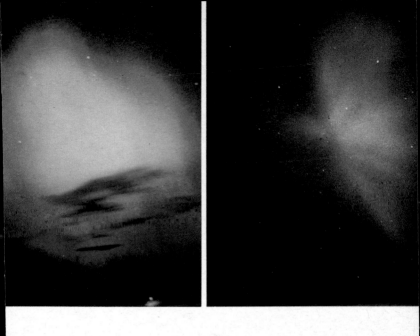

Fig. 38 **Aurora Borealis:** Northern Lights in action.

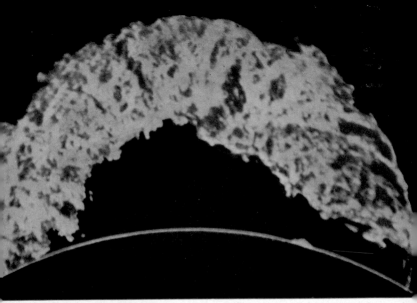

Fig. 39 Giant solar prominence.

Fig. 40 Noctilucent clouds.

Fig. 41 *Book of Hours*: Duc de Berry's early fifteenth-century calendar prayer book depicting the month of September.

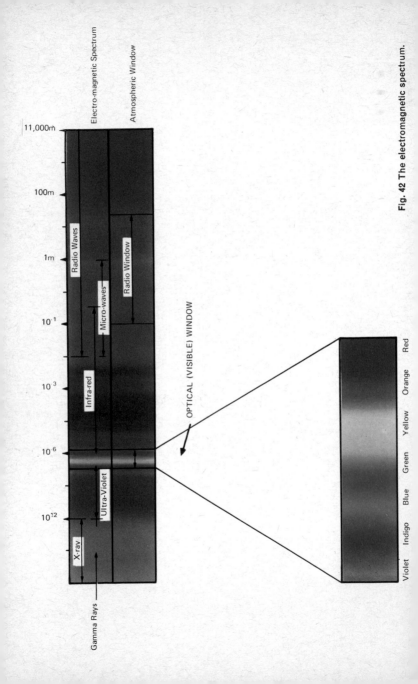

Fig. 42 The electromagnetic spectrum.

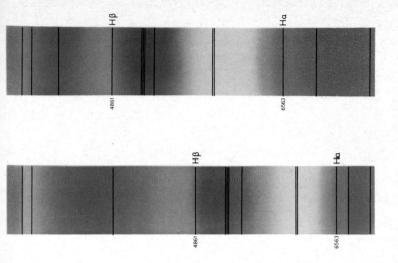

Fig. 43 Principal Fraunhofer lines in the solar spectrum: *(top)* with a prism spectroscope, *(bottom)* with a grating spectroscope.

Fig. 44 Safe method of projecting the Sun's image to view sunspots.

Fig. 45 The Solar System:
S = Sun (at centre)   S = Saturn
M = Mercury   U = Uranus
V = Venus   N = Neptune
E = Earth   P = Pluto
M = Mars   A = Asteroids
C = Comets   (minor planets)
J = Jupiter

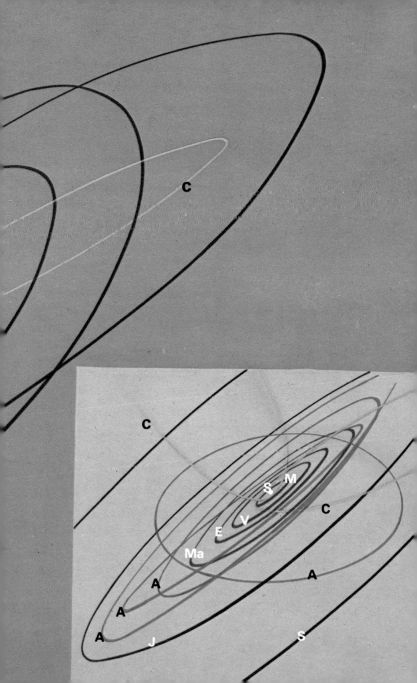

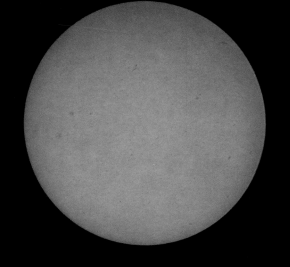

Fig. 46 The Sun.

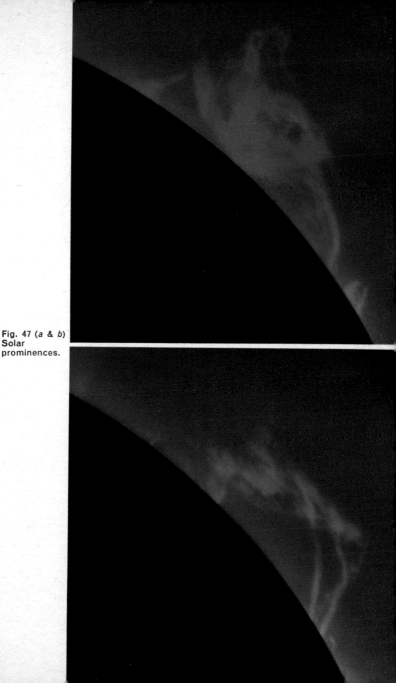

Fig. 47 (a & b)
Solar
prominences.

Fig. 48 The lunar far-side.

Fig. 49 Hadley (Delta) Mountain (near Apennine Mts) showing indications of stratification of lunar rocks.

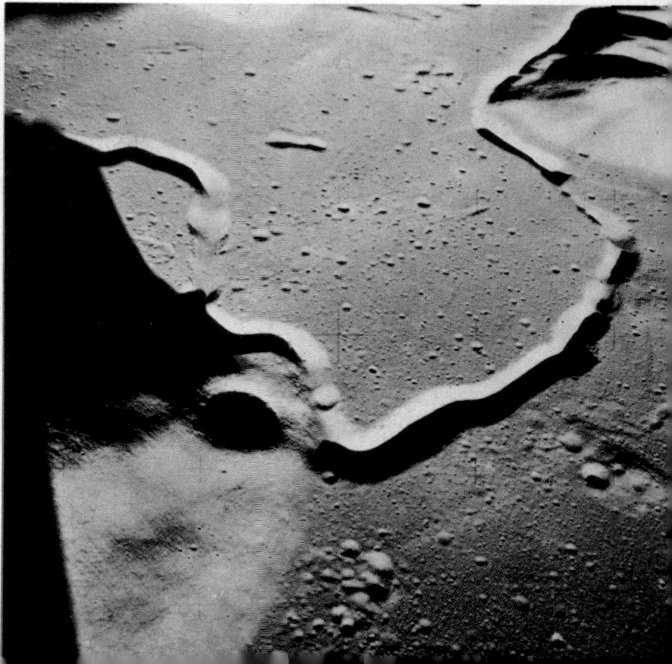

Fig. 50 Hadley Rille, a huge, deep shadow-filled 'valley' on the lunar surface 300 metres (1000 ft) deep.

Fig. 51 Aristarchus Plateau in early morning light; all the lunar surface features are sharply defined by black shadows.

Fig. 53 *(opposite):* Lunar observatory of the future.

Fig. 52 Typical densely cratered lunar highland scenery.

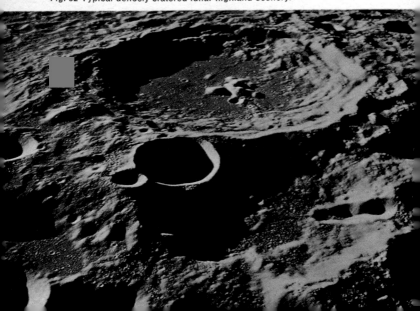

Fig. 54 Total eclipse of the Sun, Siberia, 22 September 1968.

Fig. 55 The diamond ring effect.

Fig. 56 Sun totally eclipsed. Note small prominences.

Fig. 57 Partial eclipse of the Sun.

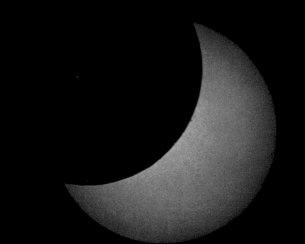

Fig. 58 Total eclipse of the Moon.

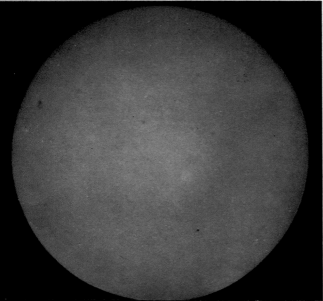

Fig. 59 Transit of Mercury across the Sun (Mercury is tiny disc, lower right).

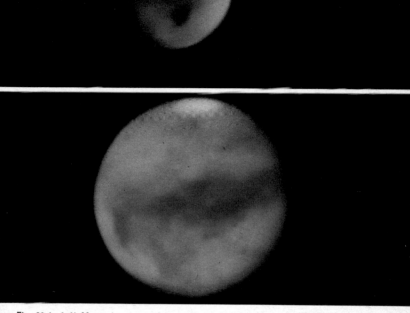

Fig. 60 (*a* & *b*) Mars showing polar caps and dark markings. (*a*) Mt. Wilson and Palomar Obs. (*b*) Lunar and Planetary Laboratory, Arizona.

Fig. 61 Map of Mars showing principal features (south uppermost).

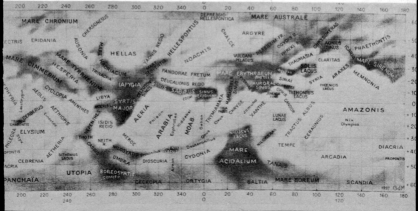

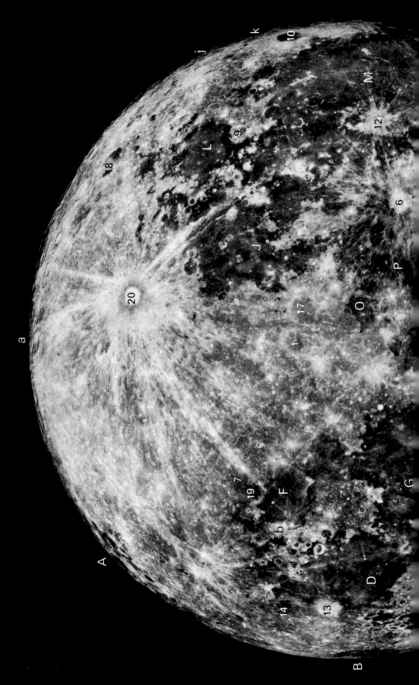

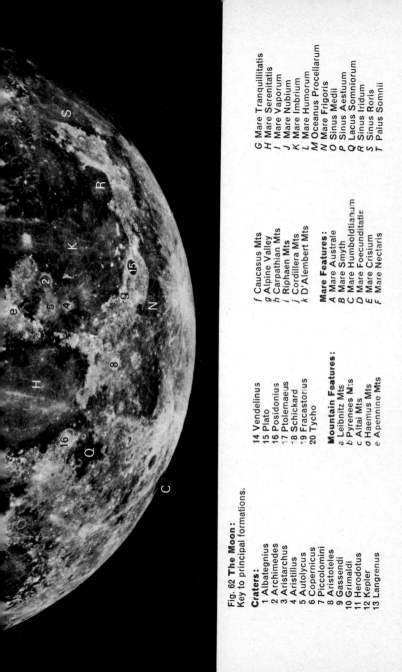

Fig. 62 The Moon:
Key to principal formations.

Craters:

1 Albategnius
2 Archimedes
3 Aristarchus
4 Aristillus
5 Autolycus
6 Copernicus
7 Piccolomini
8 Aristoteles
9 Gassendi
10 Grimaldi
11 Herodotus
12 Kepler
13 Langrenus
14 Vendelinus
15 Plato
16 Posidonius
17 Ptolemaeus
18 Schickard
19 Fracastorius
20 Tycho

Mountain Features:

a Leibnitz Mts
b Pyrenees Mts
c Altai Mts
d Haemus Mts
e Apennine Mts
f Caucasus Mts
g Alpine Valley
h Carpathian Mts
i Riphaen Mts
j Cordillera Mts
k D'Alembert Mts

Mare Features:

A Mare Australe
B Mare Smyth
C Mare Humboldtianum
D Mare Foecunditatis
E Mare Crisium
F Mare Nectaris
G Mare Tranquillitatis
H Mare Serenitatis
J Mare Vaporum
J Mare Nubium
K Mare Imbrium
L Mare Humorum
M Oceanus Procellarum
N Mare Frigoris
O Sinus Medii
P Sinus Aestuum
Q Lacus Somniorum
R Sinus Iridum
S Sinus Roris
T Palus Somnii

Fig. 63 (*a & b*) Jupiter showing cloud belts and the 'Red' Spot *(top left quadrant)*.

Fig. 64 (*a* & *b*) Saturn showing different openings of ring system (see pp. 81–2).

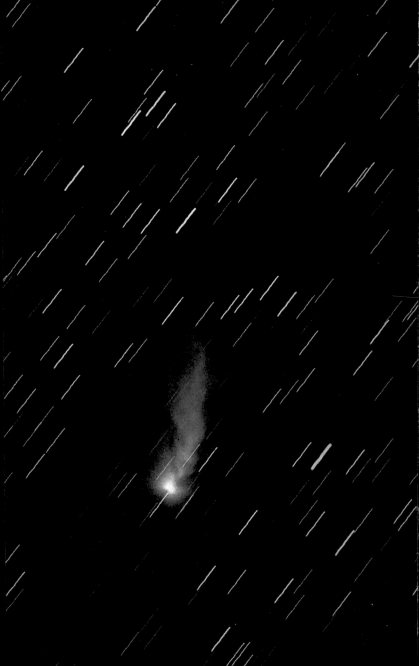

Fig. 65 *(opposite)*: Comet Humason 1961e. This large comet had a very unusual contorted tail visible from the distance of Jupiter's orbit.

Fig. 66 *(above)*: Halley's Comet in 1066 depicted in the Bayeux Tapestry.

Fig. 67 *(below left)*: Contemporary impression of the Great Comet of 1843 over Paris (the three stars of Orion's belt above).

Fig. 68 *(below right)*: Contemporary impression of Donati's Comet over Paris in 1858 (the star Arcturus near head).

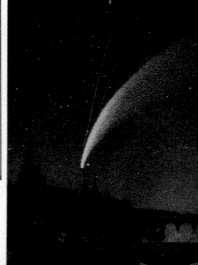

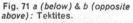

Fig. 69
Sungrazing
Comet Ikeya-
Seki 1965f in
dawn sky.

Fig. 70 Genuine and pseudo
meteorites. *Top left*, three
fragments of the Barwell
stony meteorite. *Top right*,
terrestrial iron. *Bottom*, fur-
nace slag.

Fig. 71 *a (below)* & *b (opposite
above)*: Tektites.

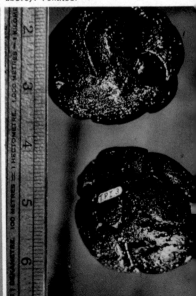

Fig. 71b

Fig. 72 Iron meteorite section showing characteristic Widmanstätten structure or pattern.

Fig. 73 Arizona Meteorite Crater.

Fig. 74 Inside the Arizona Meteorite Crater.

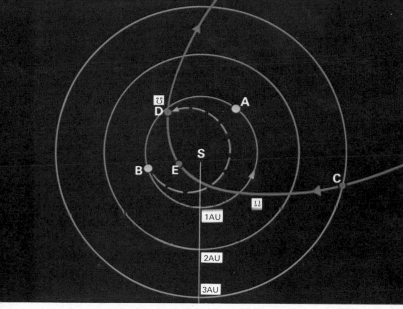

Fig. 75 Proposed probe to Halley's Comet 1985–6. Key: A & D, Earth and Comet at time of fly-by, March 1986; B & C, Earth and Comet at probe launch date, July 1985; E, Comet at perihelion, 6 February 1986.

Fig. 76 Proposed long-range planetary probes to fly-by Saturn, Jupiter, Neptune and Pluto.

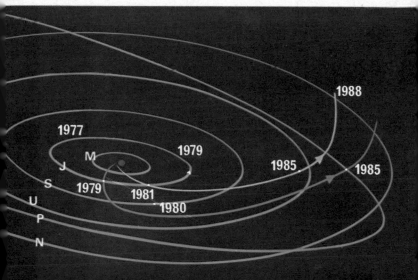

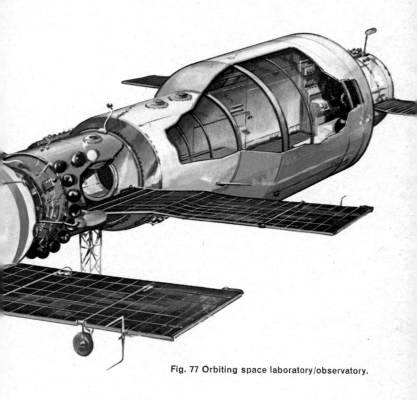

Fig. 77 Orbiting space laboratory/observatory.

a)

b)

Fig. 78 **Astronomical postage stamps.**
(*a*) 150th anniversary of the Royal Astronomical Society depicting William Herschel's 4-ft reflector. Herschel is figure on left with chart of Uranus showing satellite orbits.
(*b*) Tycho Brahe's observatory, Hven, Denmark. The supernova of 1572, and a large quadrant Tycho used for his observations.

(*c*) Galileo and his first two telescopes.
(*d*) Isaac Newton and his first reflecting telescope, and the famous apple which in dropping from the tree supposedly provided him with the ideas about gravitation.

c)

d)

Fig. 79 Planisphere dated 1702 depicting the traditional constellation figures.

Fig. 80 Part of the Zodiac of Backer dated 1680 with traditional figures reversed.

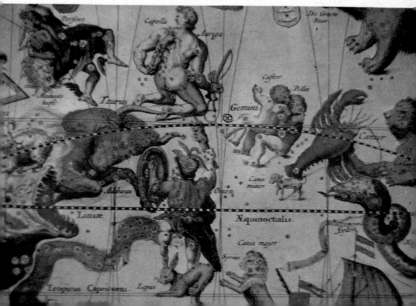

Fig. 81 **Zodiacal postage stamps:** The twelve zodiacal constellations and their signs with principal stars named. (For English constellation spellings see Appendix 4.)

BILANCIA — KIFFA BOREALIS — KIFFA AUSTRALIS — REP. S. MARINO L. 15

SCORPIONE — ANTARES — REP. S. MARINO L. 20

SAGITTARIO — NUNKI — RAUS AUSTRALIS — REP. S. MARINO L. 70

CAPRICORNO — ALGEDI — D ALGEDI — REP. S. MARINO L. 90

ACQUARIO — SADALSUUD — REP. S. MARINO L. 100

PESCI — ALGENIS — ALRISCHA — REP. S. MARINO L. 180

Fig. 82 Babylonian boundary stone or marker depicting deities in the form of animated constellation figures.

Fig. 83 Egyptian coffin lid with sky goddess Nu (or Nut) encircled by the Twelve Signs of the Zodiac and the boats of the Sun and the Moon.

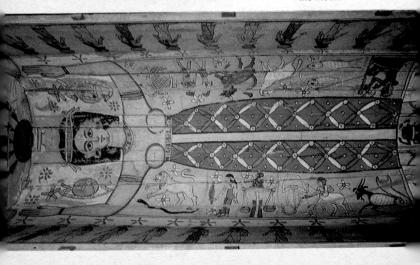

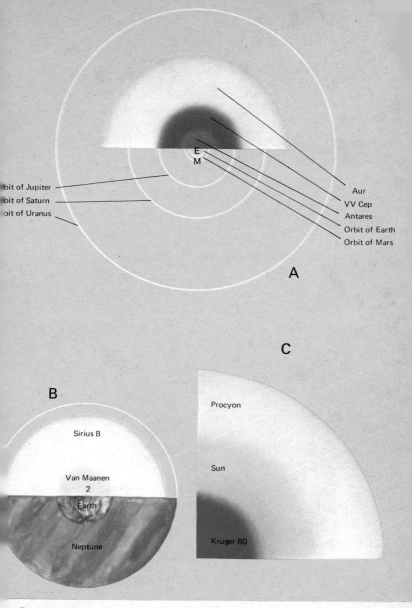

Orbit of Jupiter
Orbit of Saturn
Orbit of Uranus

E
M

Aur
VV Cep
Antares
Orbit of Earth
Orbit of Mars

A

C

B

Procyon

Sirius B

Sun

Van Maanen
2

Earth

Neptune

Kruger 60

Fig. 84 Star sizes:
(a) Supergiants compared with the diameter of the planetary orbits.
(b) White dwarfs compared to Earth and Neptune.
(c) Main sequence stars in relation to the Sun.

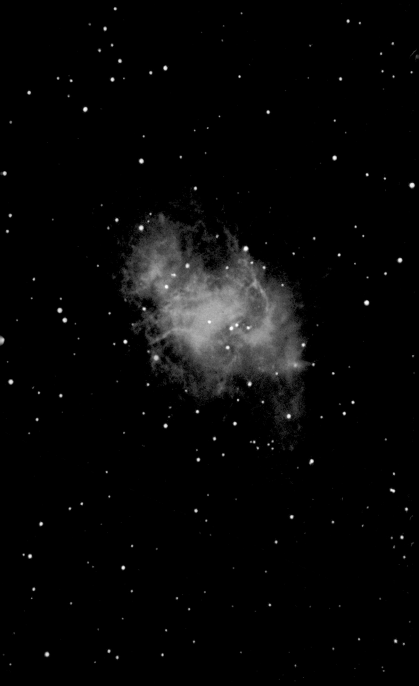

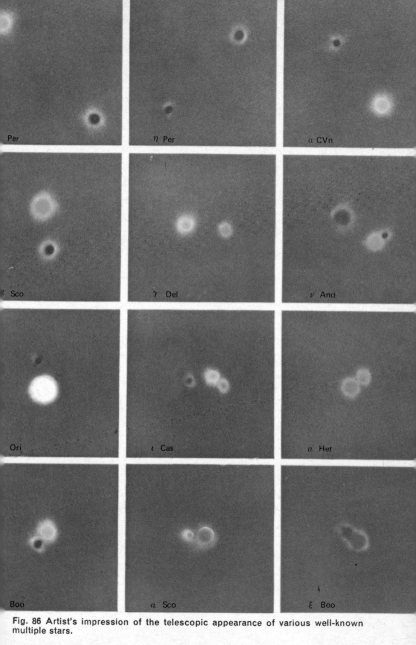

Fig. 86 Artist's impression of the telescopic appearance of various well-known multiple stars.

Fig. 85 *(opposite):* The Crab Nebula (M1) in Taurus.

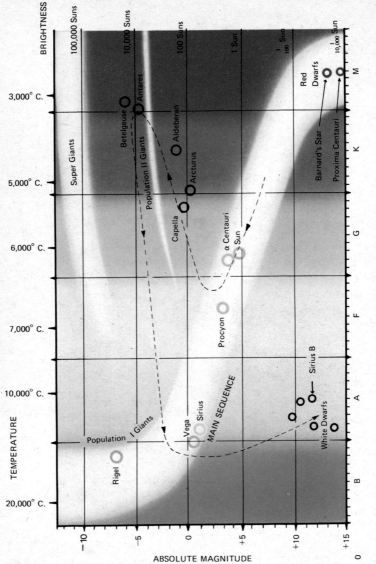

Fig. 87 The Hertzsprung-Russel diagram. The pecked line shows the *possible* future evolutionary track of the Sun.

Fig. 88 The northern circumpolar stars appearing as photographic trails owing to the Earth's rotation. Pole star is bright blob located slightly off-centre.

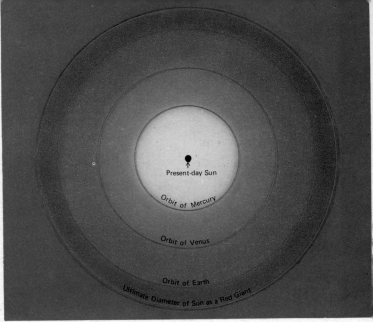

Fig. 89 One day in the distant future the Sun may expand into a Red Giant and engulf the Earth.

Fig. 90 'Black Hole' in space (see p. 218).

Fig. 91 The Orion Nebula (M42).

Fig. 92 Star Cloud in Sagittarius.

Fig. 93 Star clouds in the centre of the Galaxy (Milky Way). (The streak is due to an artificial satellite.)

Fig. 94 The Andromeda spiral galaxy (M31).

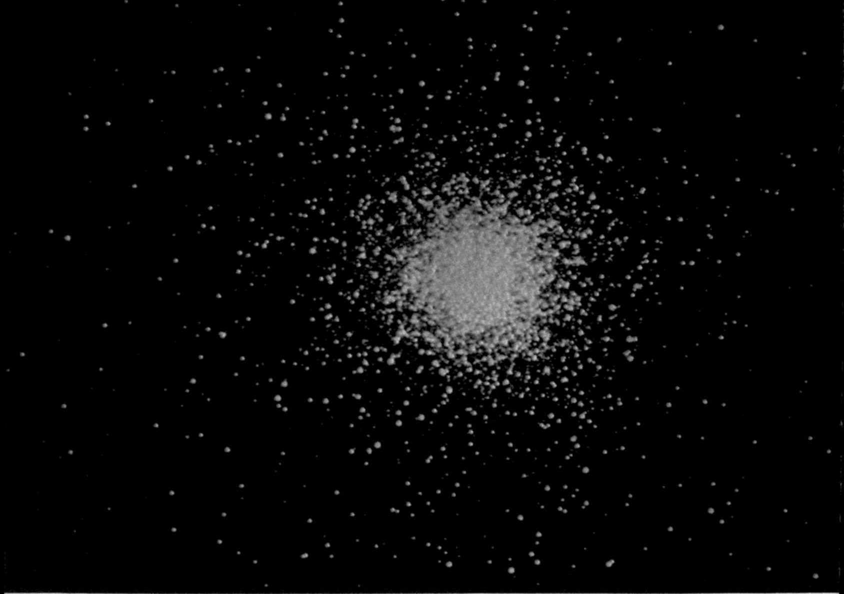

Fig. 95 Globular star cluster (M13) in Hercules.

Fig. 97 *(opposite):* The Pleiades star cluster (M45) (also see p. 220).

Fig. 96 Andromeda spiral galaxy (M31) rising above the horizon, (see Fig. 94).

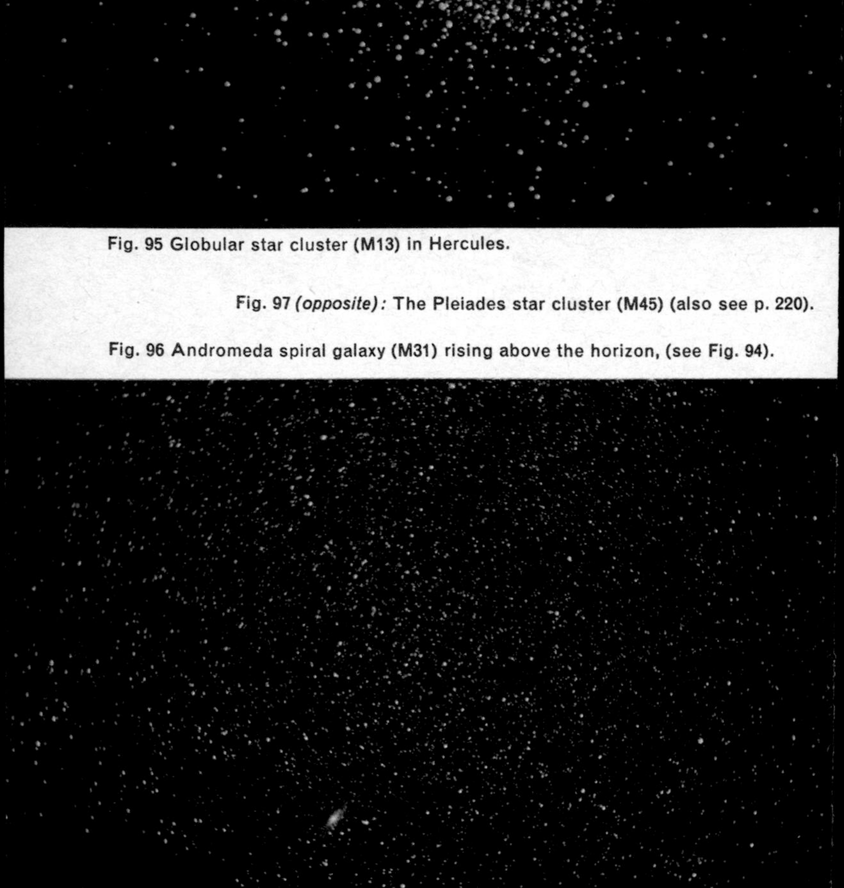

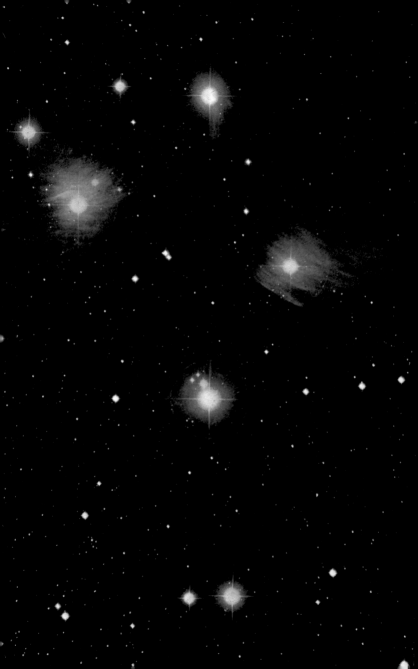

Fig. 98 The Horsehead Nebula (NGC 2024) in Orion.

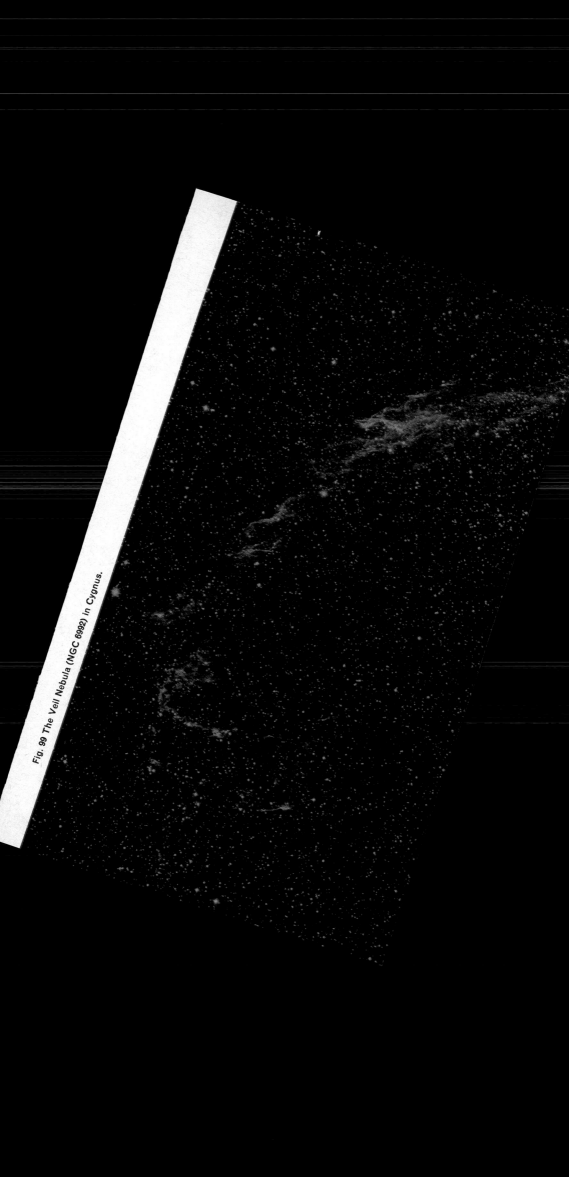

Fig. 99 The Veil Nebula (NGC 6992) in Cygnus.

Fig. 100 (opposite): The North America Nebula (NGC 7000) in Cygnus.

Fig. 101 The Trifid Nebula (NGC 6514) in Sagittarius.

Fig. 102 Gaseous nebula (M16) in Serpens.

Fig. 103 The Omega Nebula (M17) in Sagittarius.

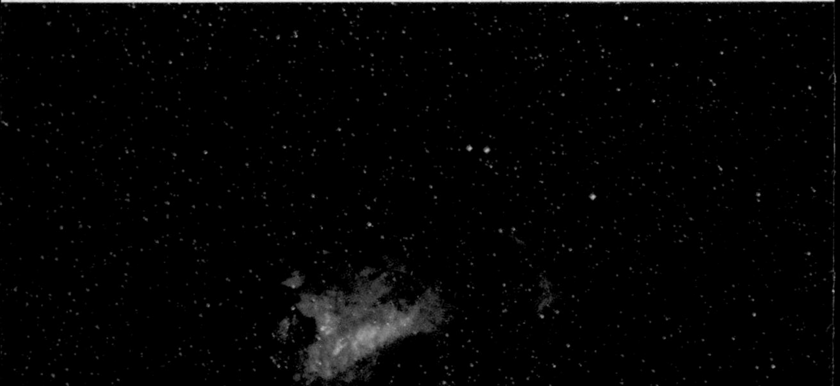

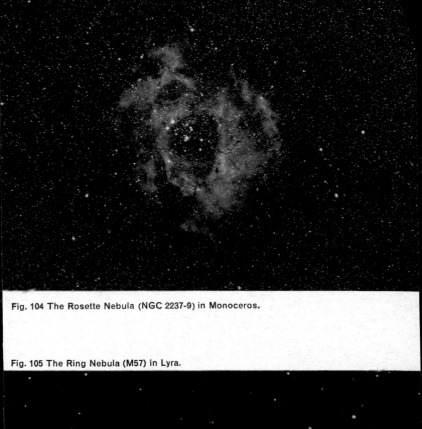

Fig. 104 The Rosette Nebula (NGC 2237-9) in Monoceros.

Fig. 105 The Ring Nebula (M57) in Lyra.

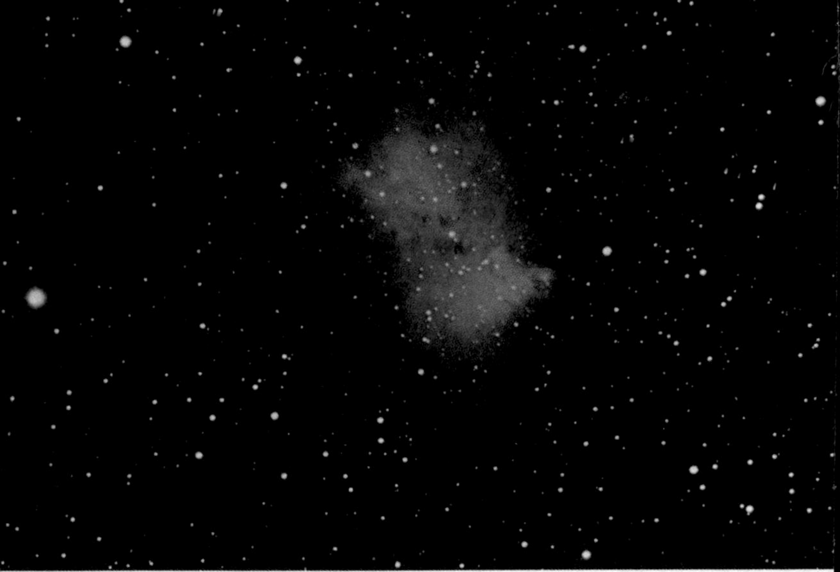

Fig. 106 The Dumb-bell Nebula (M27) in Vulpecula.

Fig. 107 Planetary Nebula (NGC 6781) in Aquila.

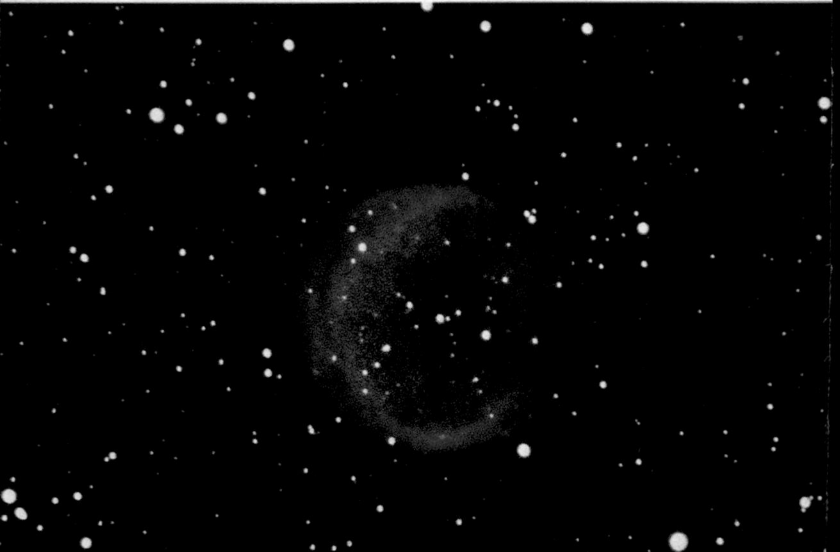

Fig. 108 Planetary Nebula (NGC 7293) in Aquarius.

Fig. 109 Planetary Nebula NGC 7685.

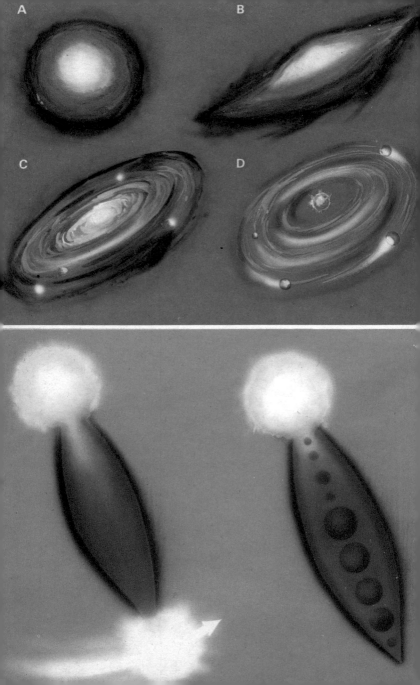

Origin of the Solar System:
Fig. 110 *(opposite top):* According to Kant and Laplace.

Fig. 111 *(opposite bottom):* According to passing star theory.

Fig. 112 According to the ideas of von Weizsäcker *(inset right:* the formation of a planet).

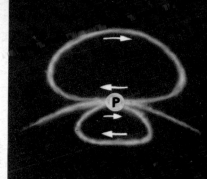

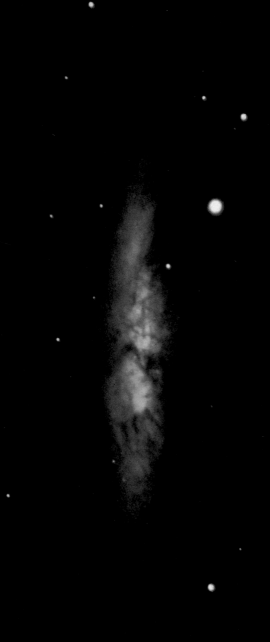

Fig. 113 Irregular galaxy (M82) in Ursa Major.

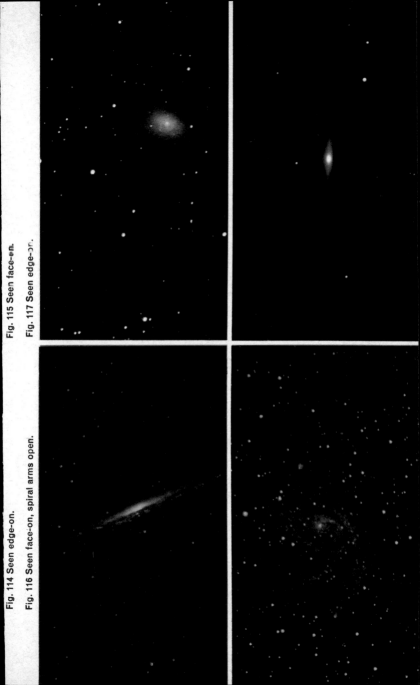

Fig. 114 Seen edge-on.

Fig. 115 Seen face-on.

Fig. 116 Seen face-on, spiral arms open.

Fig. 117 Seen edge-on.

Fig. 118 Spiral galaxy (NGC 253) in Sculptor.

Fig. 119 Spiral galaxy (M33) in Triangulum.

Spiral Galaxies: Fig. 121 Edge-on.

Fig. 122 Face-on.

Fig. 120 Whirlpool Galaxy (M51) in Canes Venatici.

Fig. 123 Nearly edge-on.

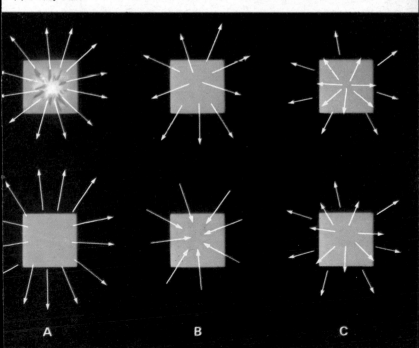

Fig. 124 Spiral galaxy (NGC 7331) in Pegasus.

Fig. 125 **Origin of the Universe** (see p. 230).
(a) 'Big Bang'.
(b) Oscillating Universe.
(c) Steady-State.

A   B   C

Fig. 126 The 48-in Schmidt telescope at Mount Palomar.

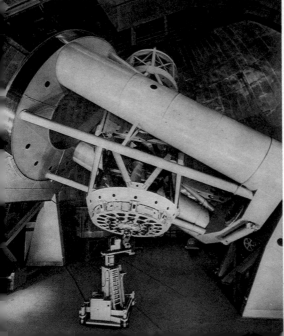

Fig. 127 The 200-in reflecting Hale telescope at Mount Palomar.

Fig. 128 The 210-ft Parkes radio telescope, N.S.W., Australia.

Fig. 130 The 120-in reflecting telescope at Lick Observatory.

Fig. 129 Control room for Parkes radio telescope.

Reference for Figs. 131–5.

Star Maps pp. 182–91.

The maps show all 88 constellations and over 1600 naked-eye stars to the limit of the 5th magnitude for the Epoch 1950. Also included are double stars, variable stars, star clusters, bright nebulae and galaxies, which may be observed with small telescopes and binoculars; see lists of selected objects in Appendices pp. 242–53.

Instructions for using star maps

The calendar months printed on the bottom of the maps (or round the edge of the circle on the circumpolar maps) show the time of year when the constellations located immediately above the month are at their highest point on the observer's meridian (or true north–south line) at 9 p.m. (2100 hours) local time. If observations are made before or after this time, remember that for each hour of clock time the star sphere is displaced one hour of Right Ascension or 15° angular measure (i.e. 24 hours = 360°).

The maps are suitable for observation *anywhere in the world*, but observers in the temperate zones of the northern and southern hemispheres will see only their own circumpolar constellations plus some of the equatorial ones. What constellations are circumpolar (or always above the observer's horizon) and what constellations an observer can see on the Equatorial Maps will depend on his latitude, and also on the time of night and the month of year. The three Equatorial Star Maps, pp. 184–89 are for use facing *away* from either the north or south celestial pole. The Circumpolar Star Maps, pp. 182–83 and pp. 190–91 are for use facing *toward* the pole of their respective hemisphere. Note that observers in the southern hemisphere should use the Equatorial Maps reversed, bottom to top.

As a general guide, observers in North America and Europe in temperate latitudes will find the north celestial pole (Pole Star) a little above the position half way between the northern horizon and zenith.

For observers who wish to find (and perhaps mark in pencil) the limit of constellation and star visibility for their own latitude, apply the following simple method:

Equatorial Star Maps: Take the latitude away from 90° and the remainder is the southern or northern declination limit.

Example: *northern hemisphere*, observer located at 51° North latitude. 90° minus 51° = 39° minus (—) declination limit.

Continued on page 190

Variable Star S^v

Cluster $*$ M103

Galaxy or Nebula O   M31

Meteor Shower

Constellation boundaries ——

Star magnitudes

0 1 2 3 4 5

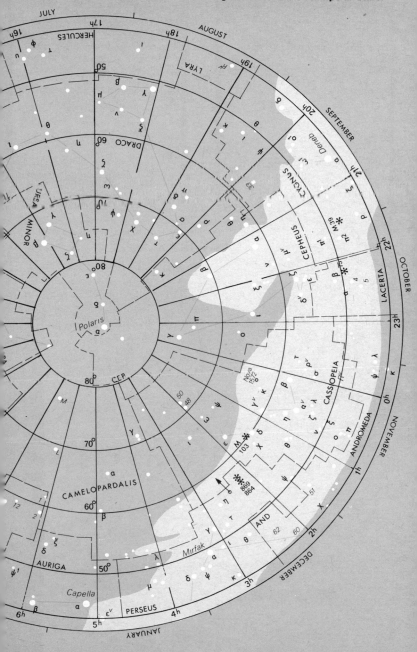

Fig. 131 Northern Circumpolar Stars.

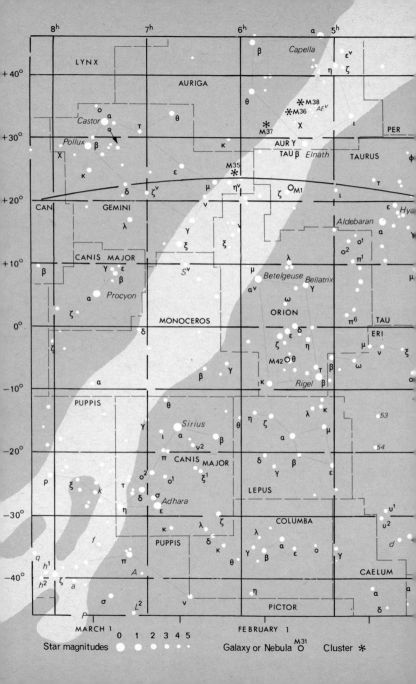

Star magnitudes ● ● ● ● ● ● 0 1 2 3 4 5   Galaxy or Nebula O ^{M31} Cluster ✱

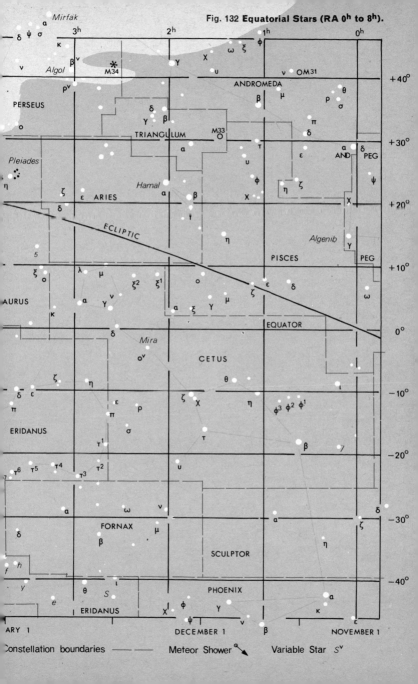

Fig. 132 **Equatorial Stars (RA 0ʰ to 8ʰ).**

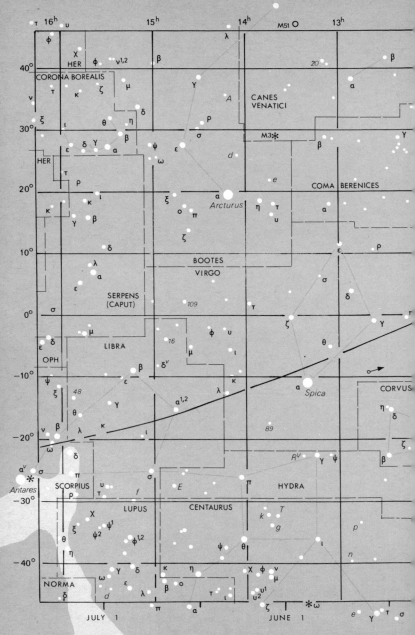

For legend see fig. 131.

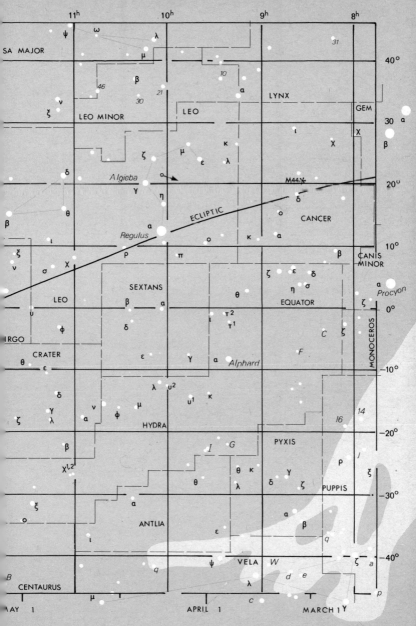

Fig. 133 Equatorial Stars (RA 8ʰ to 16ʰ).

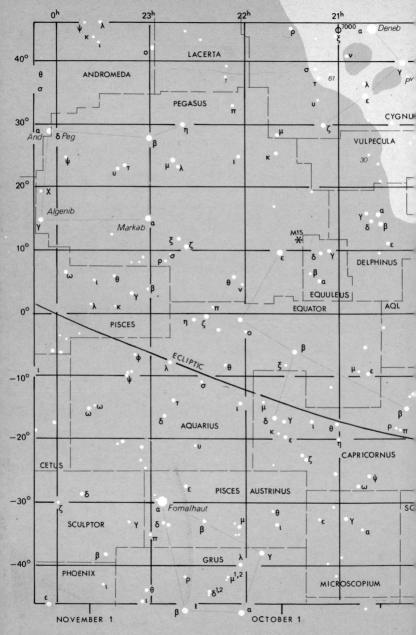

For legend see fig. 131.

Fig. 134 Equatorial Stars (RA 16ʰ to 0ʰ).

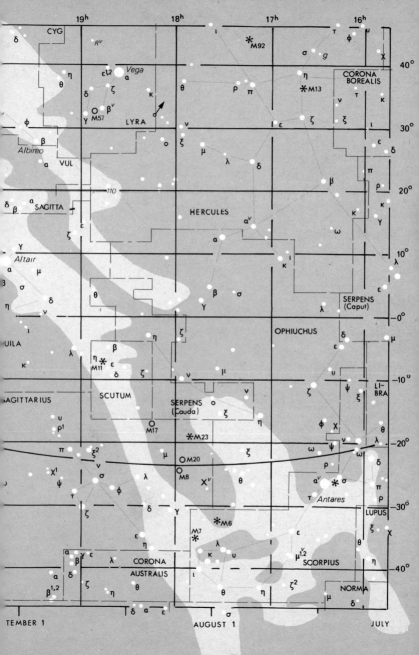

Example: *southern hemisphere*, observer located at 40° South latitude. 90° minus 40°=50° plus (+) declination limit.

Note: unlike terrestrial latitude which is measured north or south of the equator, *declination* is measured from the *celestial equator* plus (+) in the northern celestial hemisphere and minus (−) in the southern celestial hemisphere.

Circumpolar Star Maps: To determine what stars and constellations are circumpolar for a particular latitude, take the latitude *away* from 90°.

Example: observer located at 51° North latitude.
90° minus 51°=39°.

Thus at this latitude all stars and constellations in the northern celestial hemisphere with declination *greater* than +39° will be circumpolar.

It must also be borne in mind that because of horizon haziness, an observer (except in unusual atmospheric conditions) will not see stars until they have risen a few degrees above the horizon.

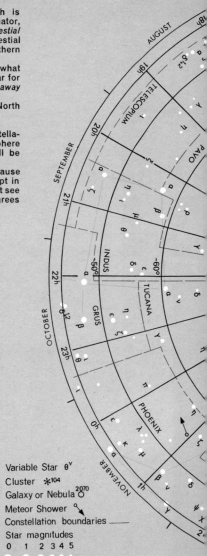

Variable Star θ^v
Cluster ✱104
Galaxy or Nebula 2070 Ⓞ
Meteor Shower °↘
Constellation boundaries ——
Star magnitudes
0 1 2 3 4 5

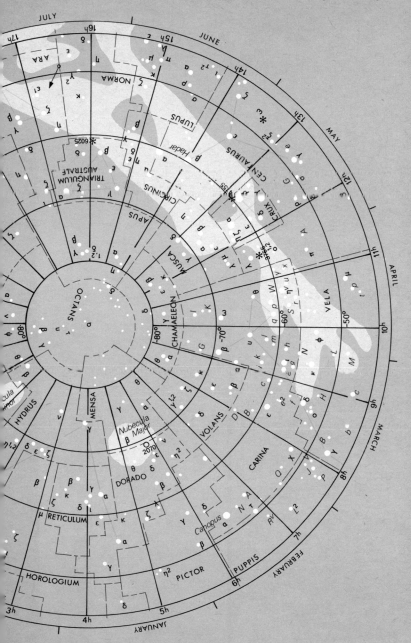

Fig. 135 **Southern Circumpolar Stars.**

Fig. 136 *(top left):* Binocular telescope 25 × 105 used for comet hunting. For comparison with instruments in Fig. 137 note size of opera glasses.

Fig. 137 *(top right):* A 2-in refracting telescope with various types of prismatic binocular and opera glasses suitable for amateur stargazing.

Fig. 138 *(below):* A Questar (catadioptric) telescope. A very compact, portable, high-performance telescope used either as a visual or photographic instrument. (Observer is H.C. Courten, Adelphi-Grumman Siberia Solar Eclipse Expedition 1968.)

Fig. 139 Total eclipse of the Sun which occurred in Siberia 22 September 1968 showing Baily's Beads as irregularities (see pp. 63–4).

Fig. 141 Asteroid (minor planet) trail (see p. 75).

Fig. 140 Crescent phase of Venus between Greatest Elongation and Inferior Conjunction.

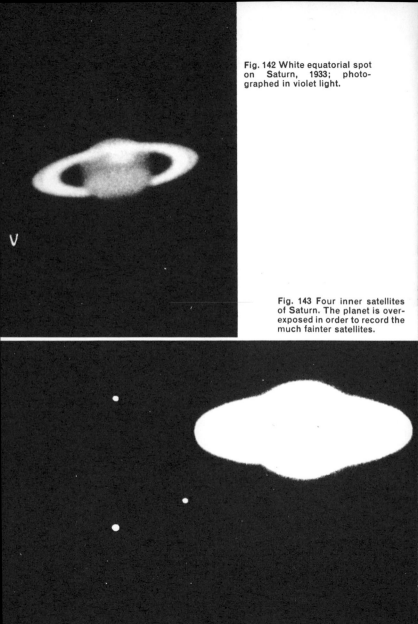

Fig. 142 White equatorial spot on Saturn, 1933; photographed in violet light.

V

Fig. 143 Four inner satellites of Saturn. The planet is overexposed in order to record the much fainter satellites.

Fig. 144 Pluto photographed shortly after discovery in 1930 (a) 2 March 1930, (b) 5 March 1930.

Fig. 145 Comet Arend-Roland
1957h(III) 24 April 1957.

Fig. 146 Comet Mrkos 1957d(V)
5 August 1957.

Fig. 147 The reassembled fragments of the Barwell (Leicestershire, England)
Meteorite which fell 24 December 1965.

Fig. 148 The Wolf Creek Crater, the largest meteorite crater in Australia, located 96 km (60 miles) south-west of Hall's Creek, W.A. It is 850 metres in diameter with a perfectly level floor on which trees are growing.

Fig. 149 The great fireball of 24 March 1933 above New Mexico. Later 4 kg of meteorite material was recovered from an area 45 km (28 miles) long.

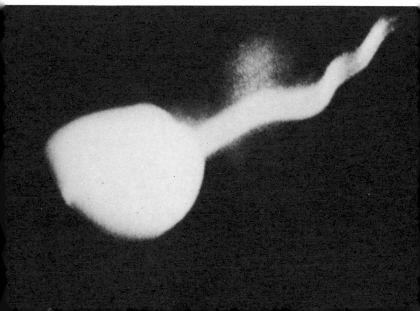

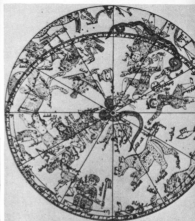

Fig. 151 The northern hemisphere of an Arab celestial globe by Mohammed ben Helal, 1275.

Fig. 150 The Atlante Farnese celestial globe. This Roman sculpture of about 200 B.C. is the first complete picture of the traditional constellation figures.

Fig. 152 Nova (HR) Delphini (arrow), an exploding 'new' star discovered by the British amateur G.E.D. Alcock 8 July 1967. The diamond-shaped pattern of Delphinus can be seen at the bottom.

Fig. 153 Cluster of distant galaxies in Corona Borealis.

Fig. 154 Faint galaxies in Coma Berenices.
Note that in both pictures the hazy galaxies outnumber the foreground stars belonging to the Milky Way.

RELATION BETWEEN RED-SHIFT AND DISTANCE FOR EXTRAGALACTIC NEBULAE

CLUSTER NEBULA IN	DISTANCE IN LIGHT-YEARS	RED-SHIFTS
VIRGO	7,500,000	H+K 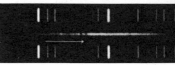 750 MILES PER SECOND
URSA MAJOR	100,000,000	 9,300 MILES PER SECOND
CORONA BOREALIS	130,000,000	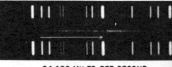 13,400 MILES PER SECOND
BOOTES	230,000,000	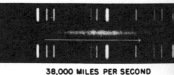 24,400 MILES PER SECOND
HYDRA	350,000,000	38,000 MILES PER SECOND

Red-shifts are expressed as velocities, c dλ/λ.
Arrows indicate shift for calcium lines H and K.
One light-year equals about 6 trillion miles,
or 6×10^{12} miles

Fig. 155 Relation of red-shift of galaxies (extragalactic nebulae) to their supposed velocity recessions.
Recently doubt has been cast on the idea that spectral velocity red-shifts are due to the Doppler effect, but the question is not yet resolved.
Distances should be multiplied by 5 to bring them into line with more recent ideas.

Fig. 156 Large Magellanic Cloud (Nubecula Major)

Fig. 157 Small Magellanic Cloud (Nubecula Minor)

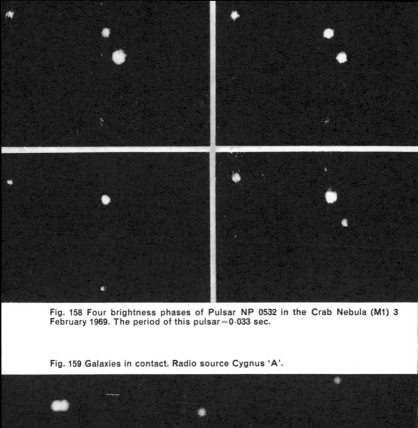

Fig. 158 Four brightness phases of Pulsar NP 0532 in the Crab Nebula (M1) 3 February 1969. The period of this pulsar = 0·033 sec.

Fig. 159 Galaxies in contact. Radio source Cygnus 'A'.

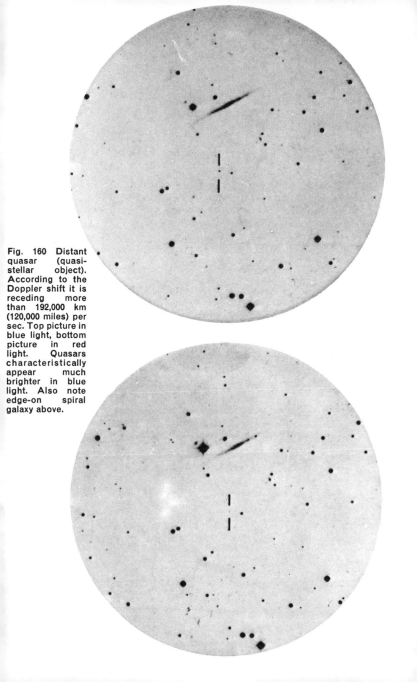

Fig. 160 Distant quasar (quasi-stellar object). According to the Doppler shift it is receding more than 192,000 km (120,000 miles) per sec. Top picture in blue light, bottom picture in red light. Quasars characteristically appear much brighter in blue light. Also note edge-on spiral galaxy above.

Fig. 161 Asteroid (minor planet) Icarus mag 13·2 (arrow) 17 June 1968. In this picture (exposure 2 mins) the telescope was made to follow the motion of Icarus so that the stars appear as trailed interrupted images.

Fig. 162 (a) New Soviet 236-in reflecting telescope during its assembly. (b) The 40-in refracting telescope of Yerkes Observatory.

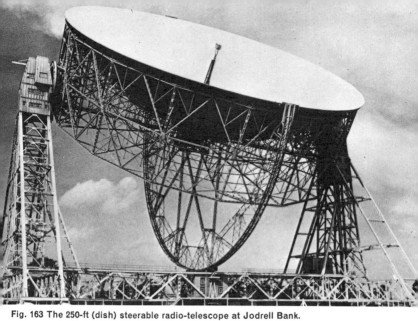

Fig. 163 The 250-ft (dish) steerable radio-telescope at Jodrell Bank.

Fig. 164 Baker-Nunn Super-Schmidt camera/telescope. This type of combined (catodioptric) telescope is constructed to operate at very short focal ratios.

Fig. 165 (a) One of the fixed mile-long arms of the Mills Cross radio-telescope near Canberra, Australia.

(b) Bird's-eye artist's impression of the Mills Cross telescope. With this kind of telescope sharper definition is achieved.

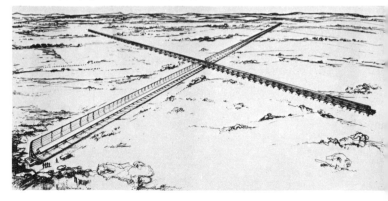

metals, and there are those with highly complex dark fluted molecular bands of carbon or titanium oxide.

When spectroscopy began in the middle of the nineteenth century, it was soon noted that star spectra fell into distinctive groups. The first classification consisted of four main groups denoted by Roman numerals I to IV (the Sun was in III). But later as spectroscopy progressed, the system had to be changed and expanded to the present-day one of using Roman capitals, thus: O – B – A – F – G – K – M in decreasing order of star surface temperature. Each main class is also divided into 10 (1–9) subgroups. The O stars at one end of the sequence are greenish or bluish-white, and the M stars are orange-red. The spectral sequence can readily be memorized using the mnemonic: Oh Be A Fine Girl and Kiss Me quick.*
In addition there are the sub-classification groups: W – R N and S. The W class precedes (), and the R – N parallels K – M. The S group also parallels M. Thus we have:

$$\text{(W)} - \text{O} - \text{B} - \text{A} - \text{F} - \text{G} \overset{\nearrow \text{R} - \text{N}}{\underset{\searrow \text{(S)}}{-}} \text{K} - \text{M}$$

The main features of the classes in term of chemical make-up and surface temperatures are as follows:

W	(Wolf-Rayet star)†	very hot, greenish-coloured stars, contain bright emission lines of hydrogen and helium	100,000 °C (?)
O	blue-white	rich in helium	35,000 °C
B	blue-white	hydrogen – helium	21,000 °C
A	white	hydrogen and weak in metals	10,000 °C
F	yellow-white	ionized metals	7200 °C
G‡	yellow	neutral metals	6000 °C
K	yellow-orange	neutral molecular bands	4700 °C
M	orange-red	molecular bands of titanium oxide	3300 °C
R N } red		stars rich in carbon	<2–3000 °C
S		like M stars and bands of zirconium oxide	

It must be remembered that the make-up of stars is highly complex and contains many features. The table above only gives their principal

* A longer variation of this is: Wow Oh Be A Fine Girl and Kiss Me Right Now or Soon.
† Very hot stars named after the two astronomers who investigated them.
‡ The Sun is a G 2 star.

characteristics in the very broadest sense. For example, the intensities of the absorption lines as we proceed from hot to cool stars is due principally to variations in temperature. It must also be borne in mind that hydrogen and helium are, by far, the most abundant substances in most stars, but the absorption lines of helium are not visible in the spectra of cool stars owing to insufficient temperature available to 'vibrate the electrons. We also find a huge variety of stars that represent transitional phases in each group.

Variable Stars

Some stars do not remain at constant brightness, and their magnitudes fluctuate either in precise recurrent cycles or at irregular intervals. Such variable stars within our own galaxy number several thousands, and there are many fainter ones yet undiscovered but in such numbers that it would be impossible to keep track of them all. The period of light variation ranges from a few hours to several hundred days. In some special instances the periods are even longer and are measured in years, particularly those stars which fluctuate at irregular intervals.

Variable stars can be classified in five main divisions (*see* Fig. 166a, b, c, d, f), excluding a group called *secular variables*, stars which during the course of several centuries have faded or brightened in comparison with the estimates made by earlier astronomers.

(1) *Long-Period Variables*. Typically orange-red or red giant stars of spectral types M and N but also includes types S, K and R. A typical star of this class is Mira (*o* Ceti), which at maximum brightness is visible to the naked eye.

The period of L.P.V. stars is between 70 and 700 days, but averages about 275 days. The cause of the brightness variation is thought to be owing to rhythmic pulsations in the outer atmospheric layers of the star.

(2) *Irregular Variables*. A group consisting of stars ranging from spectral types B to N and sometimes associated with nebulous material. Six groups are recognized.

(i) Red-tinted stars like 'the Garnet Star' (*μ* Cephei), which show minor, irregular fluctuations.

(ii) U Geminorum type. Stars which remain at constant minimum brightness for long periods then suddenly brighten several magnitudes, after which they slowly fall back to minimum light.

(iii) R Coronae Borealis type. A sub-group which remains at maximum brightness then suddenly falls to a minimum. After an unspecified period it will then suddenly rise again to maximum. The rise to maximum occurs more quickly than the fall to minimum.

(iv) RV Tauri type. These resemble the eclipsing β Lyrae variables (q.v.) but are irregular. Range is about 2 mags.

(v) Nova-like Stars. Characterized by ultra-rapid rise to maximum like true novae (q.v.).

(vi) Flare Stars. A relatively faint sub-group of stars which rise to a maximum in a very short period – often measured in minutes – and then fall back to normal brightness again.

(3) *Cepheid Variables.* These occur in two varieties.

(i) Classical Cepheids with periods ranging from 3 to 60 days.

(ii) Short-period Cepheids (or Cluster-type Cepheids), including a class known as RR Lyrae type. Periods range from 0·4 to 0·6 day. The brightness fluctuations are of small amplitude, and variations are probably due to pulsations of the star.

(4) *Eclipsing Variables.* Stars whose brightness variation is caused by a companion star revolving in common orbit and periodically eclipsing the primary. There are two sub-groups.

(i) Algol-type Stars. So named after the prototype star Algol (β Persei), the Arabs' 'Demon Star'. Characterized by a well-marked minimum and a small secondary one.

(ii) β Lyrae-type. Stars with ellipsoidal shapes owing to mutual gravitational interaction. They have two equal maxima with a small minimum, followed by a large minimum.

(5) *Novae or Temporary Stars* – colloquially known as 'new stars' or 'exploding stars'; also the 'guest stars' of the early Chinese chronicles. These stars have previously been very faint objects which suddenly rise several magnitudes (often over 20) and become extremely brilliant objects. The light output during the short period of rise may exceed 50,000 times its previous brightness. Bright novae visible to the naked eye occur once every two or three years, but telescopic novae occur more frequently.

Novae can be classified into two main groups plus minor variations within each group. The most common are ordinary novae (absolute mag -7). More rare are the supernovae (absolute mags $-13\cdot6$ to $-16\cdot5$). Only a few of the second class have been observed in our own galaxy during the past thousand years, although they are seen fairly frequently in the external galaxies owing to the huge number of such objects which are visible in large telescopes. The most famous to occur in our galaxy in modern times was observed by Tycho Brahe in 1572, in the constellation of Cassiopeia (Map, p. 183). Another previous example occurred in 1054, in Taurus, which subsequently became the Crab nebula, and recently a pulsar was also discovered there (*see* p. 217, Fig. 85, Map, p. 184). Ordinary

Fig. 166 Variable stars:

(*a*) Light curve of a long-period variable (*o* Ceti, *Mira*).
(*b*) Light curve of an irregular variable (SS Cygni).
(*c*) Light curve of a Cepheid variable (*δ* Cep).
(*d*) Light curve of an eclipsing Algol-type binary (*β* Per, *Algol*).
(*d¹*) Reason for light variations in eclipsing Algol-type binary.
(*e*) Light curve of an eclipsing *β* Lyrae-type binary (*β* Lyrae).
(*e¹*) Reason for light variations in eclipsing *β* Lyrae-type binary.
(*f*) Light curve of a nova (Nova Aql 1918).

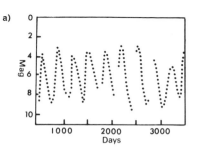

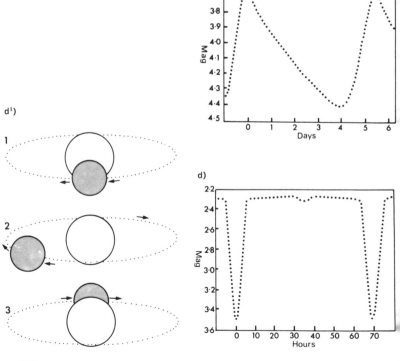

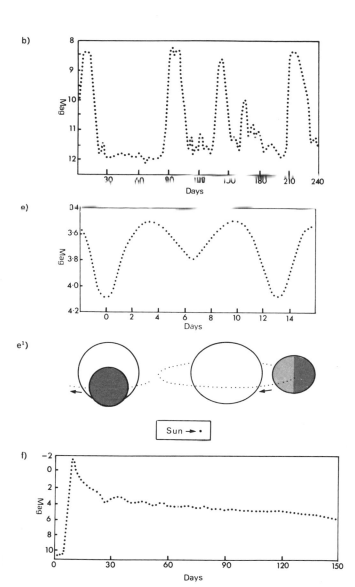

b)

e)

e¹)

Sun → •

f)

novae also consist of two main sub-groups.* Fast novae rise to maximum brightness in one or two days. Other (slow) novae take several days or several weeks.

Novae are stars which literally explode and are perhaps identical with a variety of star called pulsars, or at least they are closely related objects (*see* p. 217).

The study of variable stars provides an insight into the various physical stages which a star evolves during its lifetime. In addition the Cepheid (pronounced 'sefid' or 'seefid') variable provides a particularly useful tool by which stellar distances can be determined of stars beyond range of ordinary parallax methods. The light fluctuations of the Cepheids proceed with clockwork regularity. When a number of Cepheids had been discovered, a peculiar relationship was noticed between the length of the period of light variations and the luminosity (or intrinsic brightness): the longer the period, the more luminous the star. This period/luminosity relationship, as it was now called, has furnished a distance-finding method both in our own Milky Way and in nearby galaxies. It is only necessary to measure the period of a Cepheid in order to deduce its intrinsic brightness, and then when this brightness is compared with its *observed* brightness, the distance can be deduced. Today several thousand Cepheid-type variables are known.

A number of variable stars can be studied using only the naked eye or binoculars, and a list of such interesting variables is included in Appendix 9.

Binary Stars

Many stars which appear single to the naked eye or with low-powered instruments, appear double or multiple objects when examined with telescopes of moderate powers. The existence of double stars has been known since the seventeenth century, but it was not until William Herschel began his systematic measurements to detect parallax in the eighteenth century that a definite physical connection was established between close pairs of stars, although James Bradley had hinted at the possibility some years earlier.

There are two kinds of double star: the true pairs called *binary stars* which have a physical connection and revolve round a common orbital centre of gravity; and the *optical pairs* (or doubles) which appear double simply because, by chance, two stars lie in the observer's line of sight. These chance optical doubles may in fact be stars separated by several hundred

* There is also a class of novae called recurrent novae, such as T CrB, which has periodic outbursts, e.g. 1866 and 1946.

light years. The binary stars may form systems composed of more than two stars, and multiple groups are fairly common; sometimes there may be as many as six stars revolving round a common centre of gravity in highly stable orbits.

In many binary systems, however, the component stars are too close to be separated visually with even the largest telescopes and can only be detected, either from the anomalous motion of the apparent path of a single star, or by means of the spectroscope. When light from two discrete sources which are close together, is examined in a spectroscope, it will usually appear as a single source – that is if both discrete sources have the same directional velocity in space. If, however, they have different velocities, e.g. if one star is *receding* from the observer and the other *advancing*, then the Fraunhofer lines in the spectrum from each star do not coincide and they occur at slightly different wavelengths owing to the Doppler effect (p. 228) also (Fig. 167). Although we can never actually see the separate component stars, we can, nevertheless, determine the elements of their orbits in most cases, but this is not possible if the orbits occur exactly perpendicular to the line of sight as seen from the Earth.

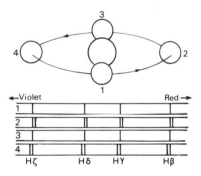

Fig. 167 Spectroscopic binary stars and the Doppler shift. The spectral (Fraunhofer) lines are duplicated when one star is receding from or approaching the observer, and are shifted towards the red or violet end of the spectrum respectively owing to the Doppler effect (*see* p. 228).

Although the closest visible double stars can only be seen in large instruments, there are many wide doubles visible with binoculars and small telescopes. One famous binary system is actually at the very borderline between naked-eye observation and optical visibility. To the naked eye Epsilon (ε) Lyrae appears as an irregular-looking mag 4·5 star located not far away from the bright star Vega (α Lyrae) (Map, p. 189). Sir William Herschel is said to have been able to split it with the naked eye, but to ordinary vision it appears only as an indistinct stellar object. The two principal stars are separated by $3\frac{1}{2}$ minutes of arc (208″), and even with theatre glasses the two are distinctly divided into two separate pin-points. The system, however, is a remarkable quadruple assembly, for each of the brighter components is in itself a double with a companion star about 2″ distant. All four stars can be resolved with a $2\frac{1}{2}$-in telescope.

Another famous naked-eye double lies in the constellation of Ursa Major (the Great Bear). With the naked eye the star Mizar (ζ UMa) (Map, p. 182) is seen to have a mag 4 companion star which the Arabs called Alcor. This was, at one time, supposedly a test for visual acuity, although nowadays average eyes can detect it quite easily if the Moon is absent and the sky dark and clear. The system Mizar-Alcor is an optical double system, but Mizar has a closer binary companion star, which can be seen with small telescopes, plus two other spectroscopic binary companions. Alcor has also a spectroscopic binary companion which makes the group a very interesting complex of stars.

Close binary stars, when seen visually through a telescope, often appear to have brilliant contrasting colours which the nineteenth-century observers vividly described in great detail, and colours such as lilac, pale primrose, mauve, emerald green, etc. were often recorded. Nowadays we know that these descriptions are false ones and are suggested in the eye by a physiological effect to which the term 'dazzle tint' has been applied.

The visibility of a particular pair of double stars in telescopes depends on the aperture of the telescope being used. The larger the telescope aperture, the greater the light gathering power (p. 235); but a larger aperture also increases the *resolving power*, or splitting ability, of a telescope. To find the limitation of any particular instrument, a simple rule was devised by the English amateur Dawes in 1854. To find the resolving power, divide 4″·5 by the aperture (or diameter of the object glass or mirror in inches). Thus, theoretically, a 1-in telescope should divide double stars 4″·5 apart. A 2-in telescope should divide 2″·25 apart (p. 247). The practical limitations are determined by the assumption of first-class 'seeing' conditions, good quality optics, and a minimum magnification of about × 25 per inch of telescopic aperture. Most binoculars, because of their low magnifying power, will *not* split double stars according to the Dawes' limit, but they can still be used for observing the wider doubles. A list of interesting double stars which can be seen in small instruments (including binoculars) is given in Appendix 7.

Stellar Evolution

In general, stars, even including novae and variable stars, are very stable bodies. But stars are continually evolving, for they are radiating energy away into space. The time-scale of stellar evolution is very long, and although we are not able to observe directly any of the evolutionary stages, we can deduce some of them by reference to the Hertzsprung-Russel diagram (Fig. 87).

Although little is known about the process, we can infer that the birth of a star begins when contractions occur within an interstellar cloud* and produce a coalescence of material. This in turn brings about a rise in temperature and a subsequent emission of radiation. Contractions of the cloud continue until the central core is hot enough to trigger the hydrogen nuclear furnaces.

A star spends most of its time on the main sequence during the period when hydrogen is being converted into helium. Stars with low initial mass, smaller than the Sun, develop slowly. Larger mass stars burn hydrogen more quickly and therefore evolve more quickly. Thus the evolution of a star depends very much on its initial mass. When hydrogen is depleted by burning, the star moves off the main sequence into the red giant branch. From here its future is less certain, and our knowledge is only sketchy. If it is a star of higher mass, it will burn helium and at a later stage carbon. The lower-mass stars, with a substantial core of degenerate matter, may suffer cataclysmic explosions subsequent to helium ignition and become novae.

The Sun took millions of years to contract from an interstellar gas cloud and evolve to become a main sequence star. Perhaps its lifetime as a main sequence star is 10,000 million years. The Sun has already been there for half this time, and in another 5000 million years it will have exhausted its hydrogen and will evolve into a red giant. By this time its volume will have expanded far beyond its present limits and already engulfed space to the distance of the Earth's orbit (Fig. 89).

Stellar Populations

Stars can be divided into roughly two groups of objects in respect to age, and in the 1940s the system Population I and Population II was evolved.

The Population I stars occur near the Sun and in galactic clusters. Population II stars occur in globular clusters away from the plane of the Milky Way. One of the chief differences between the two groups is that Population I stars have 2 per cent by mass of heavy elements, while Population II stars have only one-hundredth of this amount.

Population I stars are considered younger than the Population II members. It is thought that the heavy elements were synthesized during the evolution of the Population II stars. Much of this material has returned to the interstellar medium from which Population I stars later formed and

* Similar to the luminous and non-luminous diffuse nebulae.

are still forming (as for example in the Orion nebula, *see* Fig. 91). It can be seen that later stars will be richer in heavy elements than were the original members of the Milky Way which condensed from a more simple basic hydrogen mix. In the Hertzsprung-Russel diagram, Population I stars form the main sequence, while Population II stars form the giants and sub-dwarfs.

In some literature five population divisions are quoted, and classifications such as Halo and Disc Populations are introduced.

Ordinary Stars

In recent years astronomers have divided stars into four fundamental categories: ordinary stars, white dwarfs, pulsars and collapsars. Among the ordinary stars is included a diverse range of size, temperature and density. The term ordinary star denotes that the object contains matter very much like that of the Sun. In ordinary stars we encounter mean densities from 14 million times *less* than the Sun in red supergiant stars, to 50 times *greater* than the Sun in red dwarfs. The corresponding diameters range from 1000 times *greater* to 0·1 times *less* than that of the Sun. Despite the great variation in make-up, the ordinary stars represent a continuous distribution of closely related objects, for throughout their interiors, or at least through a substantial part, *matter behaves as a perfect gas*. In highly evolved ordinary stars, a core of degenerate matter forms in which the electrons are no longer assigned to individual atomic particles, and perfect gas laws can no longer be applied.

White Dwarfs

These are the second class of fundamental star, representing perhaps 14 per cent of all stars in the Galaxy. They differ from ordinary stars in that their sources of nuclear energy have become totally exhausted, except in the very outermost layers, and except for the outermost layer, they consist of degenerate matter. This degenerate matter is formed because electrons are not attached to individual nuclei and therefore do not sweep out orbital territorial domains as occurs within atoms in familiar states. Consequently degenerate matter is far more compressed than ordinary matter.

The average densities of white dwarf stars are approximately a million times greater than the Sun's density, and in size perhaps only one-hundredth that of the Sun – comparable to dimensions of planetary bodies. Such characteristics create an unfamiliar world. For example, the binary companion star of Sirius is a white dwarf, called Sirius B. It is 14 magnitudes fainter than the brilliant A star, but its density is such that a cubic inch of its material would weigh one ton!

Pulsars

These strange objects were first recognized on recordings made with the radio telescope at Mullard Radio Observatory in Cambridge, England, in 1967. When the signals were studied, their origin was a great puzzle which led the Cambridge radio astronomers to lightheartedly nickname them LGMs, Little Green Men!* The signals were quite different from any previous signal monitored by the radio telescopes. At precisely every 1·33730113 seconds a burst of radiation was emitted with such regularity it rivalled the accuracy of atomic clocks. The source was a discrete one, emerging from a location in the sky where no significant star had been previously noted, indicating that an entirely new kind of body was being observed.

Since the first discovery, over 100 more pulsars have been detected by radio telescopes in both hemispheres. The remarkable feature of all these objects is their precise pulsations, which cover the time range for 0·033 to 3·75 seconds in the various examples discovered so far.

It is now believed that pulsars are rapidly rotating, highly condensed stars which possess intensely powerful magnetic fields. They are probably neutron stars, which were first theoretically predicted in the 1930s. The internal pressure is so great that the orbital electrons of atomic structures have been forced on to the proton to form neutrons. This physical process is known as beta-decay and is a common reaction at normal pressures in the decay of a nucleus by emitting an electron (β particle). Pulsars represent a third kind of fundamental star in addition to ordinary stars (e.g. the Sun) and white dwarfs.

The radiation emissions from pulsars can only be detected by radio emission frequencies, and they cannot be seen visually, with the sole exception of the Crab nebula pulsar (supernova A.D. 1054). Since their discovery, some of the pulsars have been noted to vary their clock-like regularity and are found to be running down by a factor of one hundred-millionth of a second each day. The clock-like emission ranges from 34 pulses a second for the fastest known pulsars to 1 pulse every 5 seconds for the slowest. These pulses are attributed to a super-high-speed star rotation, the period of which is the actual rate of pulsation, e.g. a rotation of 34 times per second! It is thought that all radio and X-ray radiation which pulsars emit is a direct result of this very high rotation rate which beacons the signals into space like a flashing lighthouse.

Pulsars, or neutron stars, like white dwarfs probably represent a late stage of stellar evolution when nuclear energy is no longer available, but unlike white dwarfs they are thought to be the remnants of the rare supernovae explosions. Theory has recently predicted a fourth fundamental star known as
* The first pulsar is correctly called CP 1919, i.e. Cambridge Pulsar 19ʰ 19ᵐ RA.

collapsed stars, or *collapsars*, although none have yet been identified with any certainty. However, it has been suggested that the unseen components in some binary systems (e.g. ε Aurigae and 89 Herculis) might well be collapsars. The gravitational field of such stars is so intense that it results in a gravitational self-enclosing, so that no material or particles and no light emission can escape; any passing light from nearby stars would also be trapped and pulled in. The existence of such bizarre objects can thus only be detected from the intense gravitational influence on other objects nearby. The effect of a collapsar is to create a 'black hole' in space, known as a *singularity*. Such 'black holes' are not to be confused with the 'coal sacks' of the Milky Way which are owing to cosmic dust cloud obscurations.

Infrared Stars

Infrared stars are a class of ordinary star of the red giant variety, but they are so faint in the visual wavelengths of light that only large instruments such as the 200-in Palomar telescope can record them. However, at the infrared wavelengths they are brighter than anything else in the sky except for the Sun and the Moon. Special plastic mirror telescopes have been constructed to observe them in the infrared region of the electromagnetic spectrum. They comprise a class of super-red stars, and measurements show that the temperature of their outermost layer is only 700 °C, or even lower.

The Galaxy or the Milky Way

All the stars visible in the night sky belong to a great single system of stars forming the local galaxy of which the Sun is a typical member. If the sky is observed on a clear moonless night (away from a city), it will be noted that a faint band of light appears to be concentrated in a path across certain constellations. This is what the ancient astronomers called the Milky Way, of which the Greek Aratos, in his classic astronomical poem *Phenomena*, wrote: 'that shining wheel, men call it Milk'. The Greek *gala* means milk, hence *Galaxy*. The band of light extends over more than one-tenth of the visible heavens. The majority of stars forming the Milky Way are located within a horizontal plane, but they are so far distant that we cannot see them as individual points of light with the naked eye. The milky appearance is simply the effect of myriads of faint stars merged together which form a nebulous mist.

Structurally the Galaxy is like a huge disc of stars which is concentrated towards a central hub. It contains 100,000 million stars, is 100,000 light years in diameter and 5–10,000 light years in thickness. If viewed from outside, it would probably appear very much like the famous spiral Andromeda nebula

(Fig. 94), but unfortunately the view from the Earth is obscured by vast clouds of cosmic dust so that the spiral structure is hidden. Nevertheless, this structure, or rather the structure of the part containing hydrogen gas, can be detected by radio telescopes, and a complete picture can thus be built up.

The Sun is located about two-thirds out from the centre, at a distance of 30,000 l.y. Observation verifies that the whole of the Milky Way is slowly rotating, and it has been calculated that the Sun goes round once every 225 million years. But the Galaxy is not only made up of a myriad of suns. Within the flat disc, and the aura of the disc, lies a variety of matter, including globular clusters, open (galactic) clusters, gaseous nebulae, planetary nebulae, dark nebulae, graphite grains, dust clouds, hydrogen gas and a miscellany of elementary material. In recent times radio telescopes have detected emissions from water molecules, ammonia and carbon monoxide. Although we cannot see this material, it gives off its own particular invisible message in the electromagnetic spectrum which the radio telescope receives and detects. Even if the distance between stars is vast, the intervening space is *not* completely empty, yet in terms of the most efficient vacuum we can produce in a terrestrial laboratory, it would appear so (*see* Steady State Universe, p. 230).

Globular Clusters

Among the more spectacular members of the Galaxy are the globular clusters, which consist of aggregations of some hundreds of thousands of stars concentrated towards the centre so that their separate images appear to fuse together and suggest a continuous area of light (Fig. 95).

The globular clusters are situated at great distances from the Sun. The nearest is about 20,000 l.y. away, and they vary in size between 20 and 1000 l.y. Many of the clusters appear to be slightly ellipsoidal in shape, indicating that they probably rotate. Very few globular clusters occur near the plane of the Milky Way and they tend to be concentrated in regions near the galactic poles. Many of their stars are Population II and of the red giant variety, indicating great age, and it is considered that the globular clusters were formed before the Milky Way evolved its present spiral structure. The clusters do not revolve in unison with the other stars of the Galaxy, but have separate, very eccentric elliptical orbits of high velocity.

Some of the globular clusters are within range of binoculars and small telescopes. In the northern hemisphere, M 13 in Hercules is one of the best known objects (Map, p. 189) and in the southern hemisphere, ω Centauri (Map, p. 191) (*see* also Appendix 8).

The Galactic or Open Star Clusters

In contrast to the globular clusters, the galactic clusters are open, loose aggregations of stars with little or no concentration towards the centre. Generally they consist of only a few hundred stars, which are often arranged quite haphazardly.

At the present time about a thousand clusters have been catalogued, and they are all located within the plane of the Galaxy, generally lying in the direction of the spiral arms. All take part in normal galactic rotation, but this rotation probably brings about a dispersion of the members over a long period of time measured in some hundreds of millions of years. This evidence, and the evidence of their association with nebulous material in the spiral arms, leads to the conclusion that all the members of a galactic cluster have a common origin and represent a definite step in star evolution.

The nearest open cluster is 120 l.y. distant, and the present known furthest one (in the Milky Way) is about 10,000 l.y. distant. The nearest is the Hyades cluster, many members of which are visible to the naked eye and form the distinctive 'turned over' V-shape of the constellation of Taurus (the Bull) (Map, p. 184), but not the bright star *Aldebaran* (α Tau). This group is a very open aggregation with a predominant population of cool, orange-reddish stars, type K. By contrast, the Pleiades cluster, which also

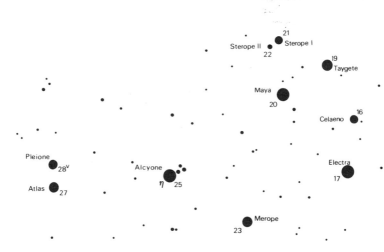

Fig. 168 The Pleiades open star cluster (M 45) (*see also* Fig. 97).

lies close by, consists of predominantly bluish-white, hot stars of type B.

The Pleiades (Fig. 168) is one of the most beautiful galactic clusters, and the best known example in the entire sky. They are recorded far back in history and are mentioned in the Bible. Practically in every ancient civilization one can note the 'Sweet influence of the Pleiades', and to many ancients they were known as the *Seven Sisters*. To ordinary eyes, only six bright stars are visible, and in many mythologies one can read various stories to account for the 'lost Pleiad'. However, observers gifted with exceptional eyesight have counted up to 14 stars with the naked eye. In theatre glasses, binoculars and small telescopes, hundreds of stars can be seen. In long-exposure photography they are recorded in thousands along with the surrounding nebulous gas which is associated with some of the hotter members (Fig. 97). This nebula* can also be seen occasionally in small 'richest field' telescopes used by comet hunters, but the sky conditions need to be very exceptional.

Around the heavens are many other fine examples of open star clusters which are visible in small instruments, and some just with the naked eye. Such an example is 'The Sword Hand of Perseus' (Map, p. 183), a double cluster that to the naked eye, on a clear night, appears like a faint misty smudge. Another such cluster is Praesepe in the constellation of Cancer (the Crab) (Map, p. 187). See also list in Appendix 8.

Galactic Nebulae

Diffuse luminous nebulae are often irregular, structureless, cloud-like wisps of luminous gas limited almost entirely to the region of the plane of the Milky Way. Typically they consist of a highly tenuous gas, chiefly hydrogen but sometimes helium, oxygen and other simple compounds such as carbon dioxide and monoxide, water and ice, methane, ammonia, carbon in the form of graphite grains or lacy spicules, plus quantities of dust. Nearly all the diffuse nebulae shine partly by reflection and partly by fluorescence. Their varied spectra reveal that the continuous background is probably due to reflection of starlight by the 'solid' or 'quasi-solid' particles and their bright emission spectra from hot stars embedded within the nebula. These hot stars of type O and B emit a great amount of ultraviolet radiation which causes the tenuous gas of the nebula to fluoresce.

Some of the luminous diffuse nebulae assume highly individual forms which have given rise to descriptive colloquial names such as: the Horse's Head nebula (Fig. 98), the Veil nebula (Fig. 99), the North America nebula (Fig. 100), the Trifid nebula (Fig. 101), etc.

The Orion nebula (or Great Nebula in Orion) has a number of very hot
* First discovered by Tempel in the mid-nineteenth century with a 4-in telescope.

221

stars embedded within it (Fig. 91). This nebula is just visible to the naked eye in the Sword Belt of Orion (θ Orionis; *see* Map, p. 184). In binoculars or small telescopes it can be seen to shine with a distinctive greenish tinge which is caused by the visible oxygen emission. In long-exposure photographs the nebula can be traced out over a very wide area of sky.

Dark diffuse nebulae are, as the name suggests, nebulae which are dark and not illuminated by embedded or nearby stars; as a consequence they appear like 'holes' or dark starless rifts running through the Milky Way (Fig. 101).

They were commented on by William Herschel during the course of his sweeps through the heavens. He thought that they represented real holes in the Milky Way and regions where stars were entirely absent. However, the researches of E. E. Barnard in 1889 showed conclusively that the 'holes' were clouds of opaque material interposed between us and the background stars of the Milky Way. Long photographic exposures show many of the dark nebulae to be faintly visible.

The dark nebulae are probably composed of the same kind of material as the luminous nebulae. They may represent an earlier evolutionary phase of material accretion just prior to star condensation taking place. Many of the dark nebulae are closely associated with the luminous nebulae, for example, 'the Horse's Head' nebula has a distinct black 'hole' (Fig. 98).

Even with the naked eye some of the dark nebulae are striking features in various parts of the Milky Way. They are often aptly described as 'coal sacks' and are best viewed with wide-angle binoculars.

Planetary nebulae were first so described by William Herschel, owing to their unique telescopic appearance which he likened to planetary discs. Most of the two-hundred odd examples are quite small and often appear like nebulous stars. Some have pronounced elliptical shapes and one of the most famous examples, the Ring nebula in Lyra, is suggestive of a smoke ring. When examined spectroscopically, they show gaseous emission-line spectra of oxygen, nitrogen plus hydrogen and helium. They represent extremely tenuous bodies that owe their visibility to illumination by a central star which in ordinary visible light shines very faintly, but which in the ultraviolet is much brighter. These central stars have surface temperatures of over 30,000 °C.

It is likely that planetary nebulae are formed as a result of surface explosions in a star (the central star) such as we observe with novae. Subsequent observations of Nova Persei 1901 reveal an expanding nebulous halo round the star which increases each year. The famous Crab nebula (M 1) is the result of the supernovae explosion of 1054, which was recorded in the Japanese and Chinese chronicles. This planetary nebula also contains a pulsar which is the only example that can be observed both visually and in the radio spectrum. The nebula has also undergone visible changes in form.

The nearest planetary nebula lies at a distance of 500 l.y. Although they are faint objects, a few can be seen in small 2- to 3-in telescopes if a sufficient magnification is employed (*see* Appendix 10).

Cosmic Rays

The existence of cosmic radiation has been known for about 60 years, but its origins are still uncertain, although its sources have in some instances been tentatively identified. When *primary* cosmic rays arrive at the Earth, they consist of a mixture of protons (hydrogen nuclei) and alpha-particles (helium nuclei) plus the nuclei of heavier elements having atomic weights of iron or greater. All these nuclei have been stripped of orbital electrons and accelerated to enormous energies.

When the primary particles plunge into the Earth's atmosphere, various nuclear interactions take place which generate a large number of electrons and form *secondary* cosmic radiation (gamma-rays and mesons). The primary radiation is the source of the background radiation which originates in the Milky Way.

The rays appear to reach the Earth from all directions, although more reach the Earth at the poles than at the equator. Their origin and source are disguised by the fact that interstellar space deflects them so that from our receiving location on Earth it is difficult to work out their original direction. However, it would appear that the origin of much of the background cosmic radiation can be attributed to supernovae explosions and pulsars.

X-ray Stars

X-rays emitted by the Sun were discovered in 1949, but it was not until 1962 that X-rays were first detected from sources outside the solar system.

Knowledge of certain astrophysical processes in stars is limited by the total absorption of certain frequencies of electromagnetic energy in the Earth's atmosphere. However, interstellar space (reckoned at a density of one atom per cm³) also becomes opaque even over distances extending to the nearest stars at ultraviolet wavelengths just short of the limiting wavelength (912 Å) which will ionize hydrogen. With decreasing wavelength, space gradually becomes more transparent, but not until the wavelength approaches 10 Å can X-ray radiation traverse the distance from the centre of the Milky Way.

To observe the X-ray sources, 1 Å to 10 Å, it is necessary to place the detector high above the Earth's atmosphere, since X-rays of the 1–10 Å band are almost totally absorbed by the residual air above 90 km. Shorter wavelength X-rays (<1 Å) are more penetrating at the altitudes reached by balloon flights (and so may be detected).

Detectors flown in rockets and in the orbiting satellites outside the atmosphere have revealed two sources of X-rays: discrete radiations and diffuse radiations. The discrete radiations may well be an entirely new class of astronomical object called *X-ray stars* with surface temperatures approaching 1,000,000 °C (compared with the Sun's 6000 °C). All these discrete sources are powerful X-rays emitters which are too weak to be observed in radio and optical wavelengths. Some of the numerous discrete sources may consist of a binary star pair held almost in contact with each other by gravitational attractions. The high temperature is likely a consequence of gaseous material being trapped between them.

One distant source of X-rays has been observed emitting from the remains of the supernova of 1054 (the Crab nebula), which is a radio and also a visible pulsar. Thus X-ray sources may be associated with supernovae remnants or even ordinary novae which occasionally and periodically fling off a shell of material from their outer surface so exposing to view the hot interior of a star for a brief interval. It is probably significant that certain of the recurrent novae involve stars in a binary system.

The diffuse sources of X-rays most likely originate from clouds of gas in galactic and intergalactic space. One such mechanism may be brought about from the collision between atoms of interstellar gas and cosmic rays. The X-rays reaching the Earth from outside the Milky Way are possibly derived from high-energy electrons which can transfer some of their energy into particles of light or photons, thus converting them into X-rays.

In recent times X-ray sources have become closely associated with black holes (p. 218). Some X-ray 'novae' may suddenly flare up within a day or so and then slowly die away over a period of several months. Other short-lived X-ray sources are represented by objects producing short bursts which last only a matter of seconds.

X-ray sources catalogued include Hercules X-1 (an optical pulsar) which pulses with a period of 1·24 seconds, and Centaurus X-3 with a period of 4·84 seconds. The system known as Cygnus X-1 – located about 20′ east of η Cygni (Fig. 134) – has provided great interest since it is suggested that it may harbour a black hole. Cygnus X-1 is coincident with a very young luminous super-giant star HD 226868 (20 times more massive than the Sun). Several of these X-ray sources display regular eclipses with periods of a few days indicating they are members of close binary systems. These systems probably involve transfer of matter from one star to a compact companion and represent close binary systems where gravitational energy is liberated in the form of X-rays.

Since the launching of X-ray satellite Uhuru in December 1970, over 100 cosmic X-ray sources have been discovered.

The Age of the Galaxy and the Solar System

There can be no doubts that the Milky Way is older than the solar system. Present estimates for the galaxy range between 8 and 9000 million years, based on the established principle that radioactive elements disintegrate spontaneously into different elements at a constant decay rate. Similar methods, applied to estimates of the age of the solar system, give an age of 4600 million years. These results are obtained by measuring the amounts of certain elements present in meteorites.

Origin of the Solar System

While it is reasonable to assume that the Milky Way evolved from a primordial gas cloud and subsequently condensed to form stars and nebulae, etc., the picture concerning the formation of the solar system is less clear.

Two general theories have been proposed: the solar system slowly evolved (like the galaxy) out of a dust cloud or nebula; or it is the end result of a cataclysmic event, either the Sun's close encounter with another star, or the result of a spontaneous ejection of material from the Sun.

The nebula theory dates back to the philosopher Kant (1724–1804), who thought that the nebula would in time form aggregations and then condense into solid planets. Laplace (1749–1827) had similar ideas except that he considered the nebula would condense more homogeneously. In modern times the theory put forward by von Weizsacker and Schmidt has generally been favoured. This idea involves the formation of vortices within the rings of a nebulous disc plus a complicated arrangement of secondary eddies leading to gradual accretion of material to form separate planets (*see* Fig. 112).

Among the cataclysmic events thought likely is that incurred by the close encounter by the Sun of another passing star which dragged out into new space a long filament of material which then began rotating round the Sun after the passing star continued on its own course. In the past this idea was looked on with favour, but in modern times objections in the form of mathematical argument have been put forward to refute it, and it now has few adherents.

IV BEYOND THE MILKY WAY

Extragalactic Nebulae

Until the 1920s it was not known for certain how large or how far the visible stellar systems extended into space. Since the eighteenth century, astronomers had puzzled over the forms of some of the telescopic nebulae which suggested the appearance of outside galaxies similar to our own Milky Way. However, it was not until 1926 that Edwin Hubble, using the 100-in telescope at Mt Wilson, was able to show that the spiral, elliptical and some of the irregular nebulae lay at distances far beyond the known boundaries of the Milky Way and were island universes in their own right, with stars, gaseous nebulae clusters, etc. like our own system.

One of the nearest of the spiral galaxies is M 31 in Andromeda (Fig. 94) which is at a distance of 675,000 parsecs (2,200,500 l.y.) and has a diameter of 50,000 parsecs, it is visible as a naked-eye object, shining as a fourth magnitude hazy smudge near the Square of Pegasus (Map, p. 185).

Classification of Galaxies

Extragalactic nebulae can be classified according to their shapes and appearances, and the usual system adopted is that first devised by Hubble during his investigations of the galaxies in the 1920s (Fig. 169). Actually, Hubble inferred that the elliptical galaxies were the youngest members on the evolutionary scale, and the open spirals the oldest, but nowadays the opposite view is held.

Hubble divided galaxies into three general categories: the ellipticals, the spirals and the irregular galaxies. Subdivisions were also introduced to account for the intermediate members of each class. Later Hubble modified the elliptical (No. 7) group and called them So. Although the Hubble system is still useful, the classifications have undergone modifications by later astronomers.

The appearance of a galaxy depends on the angle at which it is orientated in respect to the Earth. Figs. 113–124 show spiral galaxies at varying degrees of tilt.

Galaxy Clusters and Groups

Galaxies are not distributed over the sky in equal numbers. One reason is that the dust clouds in the plane of the Milky Way obscure our view into space beyond. Photographic surveys show that galaxies are concentrated towards the galactic poles where nearer galactic space is less opaque. These surveys also show that galaxies occur in clusters and groups (Fig. 153). One

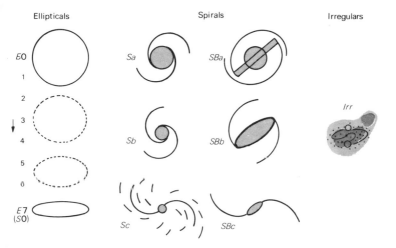

Fig. 169 The Hubble classification of galaxies.

famous group is located in the constellation of Virgo and consists of a vast supergalactic organization consisting of several hundred galaxies whose overall grouping appears to be visibly flattened. In many areas of the sky, long-exposure photographs reveal that below a certain magnitude, galaxies become more numerous than the foreground stars belonging to the Milky Way (Figs. 153, 154).

One of the more interesting assemblies is the Local Group of which the Milky Way is a member. Like the Virgo group its members occupy a volume in space like a flattened ellipsoid. It numbers about 24 examples, including spiral, irregular and elliptical nebulae. Among the irregular are the Greater and Lesser Magellanic Clouds which can only be seen in the skies of the southern hemisphere and are prominent naked-eye objects (*see* Map, p. 191). These clouds are named after Magellan, the navigator, who was among the first to describe them. Their correct Latin names are Nubecula Major and Minor, but in South Africa and elsewhere they are colloquially known as 'Cape Clouds', and in Australia, during the colonial days, as 'the Drover's Friends', since they were often used as directional aids when cattle were moved during the cool of night.

Both clouds are irregular galaxies closer than the Andromeda spiral nebula and lie at a distance of about 150,000 l.y. The larger cloud has a diameter of 25,000 l.y., and the smaller 10,000 l.y. They may perhaps be satellite galaxies or parts of the Milky Way which broke away at an earlier period of cosmological history.

In 1970 two more local galaxies were identified which were detected through the obscuring dust in the plane of the Milky Way. It is likely that other such galaxies will be found in the future, and the Local Group may be larger than at first thought.

Distances of Galaxies

All the extragalactic systems can be distinguished by their continuous spectra as would be expected from objects largely composed of innumerable stars. With large telescopes it is possible to resolve the outer edges of some of the nearer galaxies into individual stars. Among these stars we can observe novae and variable stars similar to those in the Milky Way. These include the Cepheid variables, so that when they were recognized, it was easy to apply the period/luminosity relationship (p. 110) in order to find their distance. The first estimates of the Andromeda nebula (M 31) gave a distance of about one million light years. However, later, when the 200-in Mt Palomar telescope followed up this work, several errors were discovered in the previous period/luminosity calibrations. By using observations of long-period variable stars it was concluded that all the island universes lay at twice the previously accepted distances. In the case of the Andromeda nebula, the figure now accepted is 2,200,000 l.y.

Doppler Effect

This effect was first described by C. Doppler in 1842 and has far-reaching consequences when applied to astronomy. Briefly, when a source of sound or light has a motion relative to the observer, there occurs a shift of frequency whose direction depends on whether the source is receding or approaching. In sound, the shift can be detected by a change in pitch as when an express train passes an observer. The pitch of the train is higher as it approaches but lower after passing. With light, the effect shows itself by shortening its wavelength from an approaching source and lengthening its wavelength on a receding source. Thus in an approaching source, the Fraunhofer lines will be displaced towards the blue end of the electromagnetic spectrum, and towards the red end if the source is receding. However, only motion in the line of sight of the observer can be detected in this method.

In astronomy the rotation of the Sun or a planet can be detected by examining one limb and then the other. The displacement of the Fraunhofer lines will then give the direction of rotational movement. The existence of invisible binary pairs too close to be resolved by optical telescopes can also be detected (p. 213).

One of its most important astronomical roles is in the observation of extragalactic nebulae. All except the Andromeda nebula appear to be receding from the Earth at high velocity, and the deeper one looks into extragalactic space, the faster the recessions appear to be. Hubble, in his work on galaxies, discovered a relationship between the distance of a galaxy and its *red shift*, as the recession velocity is called. The greater the measured shift of the spectral lines, the greater the distance. This relationship has had far-reaching consequences in cosmology, since it would appear from the red shift that the Universe is expanding. In the case of the quasars, the fastest recession is 283,000 km (177,000 miles) per second, which is approaching the speed of light. Doubt has been cast on the validity of the Doppler effect in interpreting such high velocities, but consensus opinion at the present time supports the idea.

Radio Galaxies

Radio galaxies were first interpreted as radio stars, but later it was realized that they represented considerably larger objects and so were renamed *radio galaxies*. All ordinary galaxies like the Andromeda nebula (M 31) and our own Milky Way give out some radio emission. However, radio galaxies emit radio wavelength radiation 14 million times greater. Many of the early type spirals and ellipticals have only a little emission, and it would appear that Population I stars (p. 215) determine the amount of emission sent out by a radio galaxy.

One of the major radio galaxy sources is known as *Cygnus A*, estimated at a distance of 300,000,000 l.y. But Cygnus A has been found to be two powerful radio sources with a visible galaxy lying between them. It is accepted that double radio sources of the kind associated with Cygnus A are formed by some violent process when energetic particles are expelled from a galaxy.

Quasars

Investigations of radio galaxies in the 1960s led to the discovery of another class of object called *quasars*, a name derived from the earlier description Quasi-Stellar Object or Source (QSO or QSS). These objects are the most remote sources of emission in the Universe, if the distances measured by the Doppler shift technique (above) are correct.

They appear to be two classes of object:

(*a*) Star-like bodies identified *with* a radio emission source.

(*b*) Star-like bodies *without* a radio emission source.

One of the most interesting features of both classes is that their emissions are variable, and large changes in the optical (visible) wavelengths occur over short periods. This provides a significant clue to their size. They cannot be large-diameter bodies, for the emitting region can be no greater than the distance light travels in the time interval over which the object varies. Variations also occur in the radio emission but to a lesser degree and over longer periods.

At present no satisfactory explanation for quasars has been put forward. If they lie at the colossal distances that have so far been measured (the furthest 7,000 million l.y.), then their emissions are inexplicable in terms of present-day physics, for the emitting power of these small-diameter objects is over 200 times greater than a giant-sized, normal galaxy.

Nature of the Universe

Cosmology is the study of the Universe as a whole including its structure, its origins and its subsequent evolution. The age of modern cosmology began after the publication in 1915 of Einstein's ideas on general relativity which broke new ground in the fundamental thinking about space and time. Hubble's work in the 1920s, with the 100-in telescope at Mt Wilson, revealed that the galaxies were apparently receding at great speed, and this was interpreted as an *expansion of the Universe* round which all cosmological theories are centred.

Origin of the Universe

Many of the theories explaining the origin of the Universe begin at a definite point in time. Among various models of the *evolutionary* Universe is the 'Big Bang' idea which considers that the Universe began with the explosion of a primeval fireball (or primordial central atom) and expanded in all directions, which it continues to do at the present time in confirmation of the red shifts we observe in the spectra of distant galaxies. In some variations of this theory the expansion is halted (or in others slowed down) by gravitational forces and then begins to contract again; at a critical stage expansion occurs again, and the pattern repeats itself in a cyclic rhythm.

The *steady-state* theory puts forward the alternative idea of the continuous creation of matter, and that large-scale properties of the Universe do not change in space or time. The expansion of the Universe is accounted for by the postulation that *new* matter is constantly being created to fill in the spaces (or voids) left. This new matter will be hydrogen, which is spontaneously formed at such a rate that it is everywhere, and the same, at all time.

Both the 'Big Bang' and the steady-state models have considerable support, and the champions of each theory periodically bring forward fresh evidence to support their cases. The 'Big Bang' adherents claim that the background microwave radiation picked up by radio telescopes represents the remnant signals emitted by helium atoms at the time when the 'Big Bang' explosion occurred, some 100 billion years back in time. The steady-state champions, however, claim that this radiation originates from distant radio galaxies or similar objects.

In the past one of the greatest obstacles facing the steady-state idea was in connection with the origin of the chemical elements. In the 'Big Bang' theory it is fairly easy to account for the nucleosynthesis of all the elements. It was claimed that in the steady-state idea there were simply not sufficient kinds of stars to manufacture them. However, more recently the steady-state workers believe they have satisfactorily accounted for nearly all the elements by way of nuclear fission processes which occur inside the super-heated cores of certain stars.

Physics of Matter

Since Einstein showed that many of the older established classical laws of physics needed drastic revision, speculation has continued about the nature of matter. Is it the same for every part of the Universe? One idea involves the coexistence of two kinds of matter, one of which is the mirror image of the other. The idea goes back to Dirac's theoretical notion in the 1930s that there should exist atomic particles with the same properties of the electron (negative electric charge) but with a positive charge. In fact the anti-proton, as it is called, was artificially created in the early 1960s in particle accelerators.

The bizarre consequences of anti-matter, or particles of anti-matter, is that it annihilates itself whenever it is coupled with corresponding ordinary matter, giving birth in the process to gamma rays and ordinary radiation. Such interaction taking place in the far corners of the Universe could account for the same kind of powerful radiation at present detected by radio telescopes.

A theoretical particle called a *tachyon*, from the Greek *tachys* meaning 'swift', has raised interesting questions concerning the ultimate speed of matter. Einstein showed that matter cannot exceed the speed of light and this is a barrier which ordinarily cannot be crossed. However, this theoretical particle, the tachyon, has some interesting properties which enable it to evade this limitation. Although at the present time no such particle has been detected, there is good reason to believe they may exist.

Even the interpretation of the Doppler effect on the recession of distant galaxies has been questioned. The red frequency (or colour) shift may be induced by other causes. Photons of light may lose energy when travelling great interstellar distances, which would cause a shift of frequency at the receiving end – comparable to the Hubble red shift.

Physicists and astronomers have long sought the 'Rosetta Stone' which would equate the microscopic world of atomic physics with the macroscopic world of the cosmos, but at the moment the finding of such a universal code appears very remote.

Origin of Life

Perhaps one of the most intriguing questions concerns the origin of life and in particular its origins on Earth. Did it evolve spontaneously on Earth, or was it brought here by a meteorite or cometary impact?

During the early 1950s, an experiment was performed by S. L. Miller in the U.S.A. under the direction of H. C. Urey. This conceived the idea of mixing together the gases ammonia, methane, water vapour and hydrogen which were assumed to be abundant in the primitive atmosphere of the Earth. The mixture was then circulated through an electric discharge, and at the end of a week, it was discovered that the water contained several types of amino acids. These substances are closely related to organic forms of life. The results of the experiment indicate that life on Earth could have occurred by the interaction of electrical energy with an elementary chemobiological 'soup'.

However, since the fall of the Orgueil meteorite in France during the nineteenth century, there has been speculation as to whether 'life spores' could reach the Earth's surface from outer space. Although the evidence for organized organic material in the Orgueil meteorite now seems very doubtful, similar carbonaceous meteorites have been found to contain organic-like hydrocarbon substances. The chief difficulty is that meteorite samples can quickly become contaminated shortly *after* falling to Earth. The Murchison carbonaceous meteorite which fell in Australia during the latter half of 1969, has been carefully investigated by many workers, and it is claimed that it does indeed contain organized organic forms which were present before impact.

Another interesting development has been the detection of formaldehyde and ammonia in interstellar space. The reaction between these two substances yields various forms of amino acids. From this evidence it would appear that the conditions for organized life may exist elsewhere in the Universe other than uniquely on the Earth's surface. Further speculation leads on to the conclusion that the Universe contains such a vast number of cosmic objects in all kinds of varied environments that it is reasonable to suppose that intelligent life could evolve independently elsewhere many times over.

V ASTRONOMICAL INSTRUMENTS

The Astronomical Telescope

The human eye is one of the most remarkable and sensitive receptors in the visible waveband of the electromagnetic spectrum and is able to *detect* a single quantum of light under suitable conditions. However, as a *collector* of light it is quite inefficient, since the pupil aperture is only a quarter of an inch (7 mm) in diameter.

Because of this technical limitation, man's knowledge of the Universe, beyond naked-eye visibility, was held in check until the invention of the telescope, which is generally attributed to the Dutch spectacle-maker Lippershey in 1608. Simple lenses had been known since ancient times, and there are numerous apocryphal stories about their different applications, such as their use as burning-glasses to set fire to enemy ships in time of war. However, the best authenticated evidence of their ancient use is the finding of a transparent lens-like artefact in Babylonia which is thought to have been employed to read cuneiform tablets.

The application of the telescope to astronomy is attributed to Galileo, who manufactured his own instrument soon after hearing the news of the Dutch discovery. However, Simon Marius and others independently applied telescopes to the heavens and were contemporary with Galileo in making the first new telescopic discoveries.

Optical Telescopes

Refractor. This is the kind of telescope first developed for astronomical use by Galileo in 1609, and although Kepler suggested modifications in 1611, which remain the basic design of present-day refractors, the Galilean optical system is still utilized in theatre glasses and inexpensive binoculars.

The refracting telescope employs an objective lens, referred to as the *object glass*, which determines the size of the instrument. Thus a 3-in telescope has an object glass 3-in in diameter. The object lens *collects* incoming rays of light from the object under view and then bends (or refracts) them to form an image behind at the *focus*. Here another lens (or lens system) is employed as an eyepiece to magnify the focused image (Fig. 170a).

Although present-day refractors are more sophisticated instruments than those employed by the earlier astronomers, the optical principles remain the same. The early refractors suffered from a serious defect known as *chromatic aberration*. A single lens of glass does not focus all colours of the spectrum at the same point. It was not until 1733 that the problem was solved when

Chester Moor Hall invented the *achromatic* lens system which almost fully corrects the chromatic aberrations. The achromatic object glass is made up of two or more lenses of different kinds of glass, usually the flint and crown variety. The combination of different varieties of glass cancels out any tendency for incoming light to disperse into the rainbow colours of the visible spectrum.

Reflector. This instrument was invented by Isaac Newton about 1670 in order to avoid the effect of chromatic aberration. The reflecting telescope employs a concave mirror instead of a lens. When light is reflected from a parabolic mirror, all the light is focused in one spot and chromatic defects are entirely absent. The mirror is usually positioned at the bottom of the tube and focuses the collected light back up the tube on to a small secondary

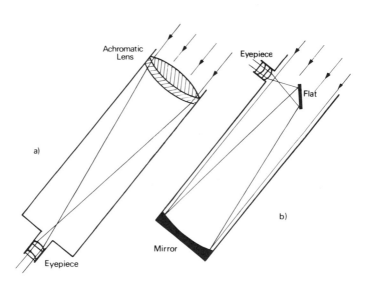

Fig. 170

(a) Principles of the astronomical refractor.
(b) Principles of the astronomical reflector.

flat mirror tilted at 45° to project the focused image sideways. It is examined by an eyepiece in a similar manner to the refractor (Fig. 170b). Like the refractor, the size of the reflecting telescope is determined by the diameter of the principal light collector, i.e. the main mirror.

Magnification, f/ratio and Resolution

The magnification of a telescope is independent of the diameter of its object glass or mirror. Magnification depends on the focal length of the object glass or mirror *divided* by the focal length of the eyepiece. Thus a telescope with a focal length of 60 in and an eyepiece of focal length 1 in gives a magnification of ×60. If the eyepiece was $\frac{1}{2}$-in focal length, the magnification would be doubled (×120).

The f/ratio, or focal ratio, of a telescope is determined by the diameter of the object glass or mirror in relation to its focal length, i.e. focal length ÷ dia. object glass or mirror. Thus a 3-in telescope with a 45-in focal length has an f/ratio of 15 (f/15). Usually refractors have f/ratios between f/10 and f/15, while reflectors are usually f/4 to f/8. However, specialized instruments of either kind are commonly met with outside these limits.

The resolving power (or resolution) of a lens or mirror is dependent on its diameter. The larger the diameter, assuming that the optical quality is first class, the greater the resolution of fine details. Resolving power is important in a telescope's ability to divide binary stars (*see* p. 214) and for observing fine planetary surface detail.

Limiting Magnitude

The ability of a telescope to reach a particular magnitude depends on its light-collecting power, i.e. the diameter of the object glass or mirror. The larger the object glass or mirror, the fainter the magnitude observed. The naked eye is, of course, a tiny telescope with an objective lens about one quarter-inch in diameter, and its limiting magnitude is about 6. With a 1-in telescope the collecting area is greatly increased and can reach mag 9 (*see* table p. 247), while the 200-in Palomar telescope reaches mag 24.

Astronomical 'Seeing'

Astronomical 'seeing' is a term used to describe the sky observing conditions in relation to the *steadiness* of the telescopic image; it does *not* describe sky *transparency* although the two are generally very closely related.

The twinkling of the stars on a cold frosty night is due to the effect of atmospheric turbulence which causes the path of the starlight beam entering the Earth's atmosphere to be displaced at rapid intervals. Any telescopic magnification greatly increases, or exaggerates, this effect, which can readily be verified if observation of a celestial object is made through a telescope or binoculars over a chimney!

The worst 'seeing' usually occurs on the breezy, frosty nights when telescopic images will often appear to 'boil'. These periods, however, produce very transparent skies which are ideal for finding faint objects such as nebulae and comets. The best 'seeing' occurs on still, misty nights when the finest views of planetary detail and double stars are obtained, but which often render the fainter objects invisible. The direct effect of 'seeing' on astronomical observation is to curtail the practical limit of magnification that can be applied to celestial bodies.

Large Optical Telescopes

Nowadays there are many varieties of astronomical telescope based on the principles of both the refractor and the reflector, and some advanced models combine both (Fig. 172). Professional observatories overwhelmingly favour the reflector over the refractor since it is easier and cheaper to construct and more convenient to use. One great advantage of the reflector is its shorter length for a given aperture, and even in long f/ratios the light path can be 'folded' back within the tube. Large refractors with f/15 ratios require large, expensive protective domes. Most of the contemporary reflectors can be adapted to operate at different f/ratios. For wide-angle photography they operate as low as f/3–4; while for use with the spectrograph (p. 239) they can be converted to f/30 or more (Fig. 171). Refractors are also limited by other design factors. The object glass has at least four optical surfaces which must be worked accurately, as against only one in reflectors (ignoring the small secondary flat). Above a certain critical size an object glass begins to distort under its own weight and cannot be supported as light must be transmitted through it; on the other hand, a reflector can be substantially supported from behind its reflecting surface. Perhaps the limit for ultimate size of a reflecting telescope is about the 300-in mark. This critical limitation is brought about by structural engineering problems rather than by optical ones. Telescope mirrors are usually constructed of low-expansion glass such as quartz or pyrex and given a long-life reflective coating of aluminium.

The largest operational refractor is the 40-in Yerkes telescope in Chicago constructed as long ago as 1898 (Fig. 161). The largest reflector is the 200-in Mt Palomar instrument (Fig. 127), but by 1975 the Soviet 238-in will be fully operational (Fig. 162). There are several 100 to 200-in telescopes, either operational at present or planned for commissioning in the near future.

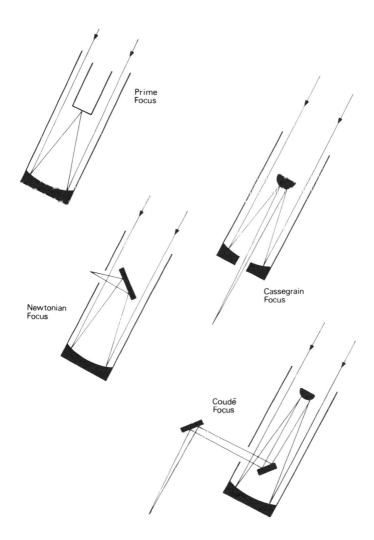

Prime
Focus

Newtonian
Focus

Cassegrain
Focus

Coudé
Focus

Fig. 171 Various alternative optical arrangements used in large reflecting telescopes.
Some instruments (Figs. 127, 130, 162) may be adapted, as required, to operate in any of
the four methods shown.

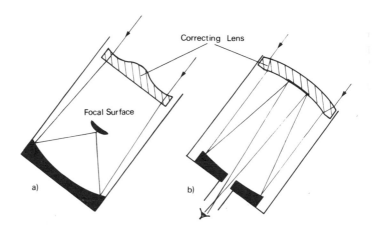

Fig. 172 Combination telescopes:

(*a*) The Schmidt telescope incorporates both refractor and reflector principles. The mirror is spherical, and the image is *first* corrected by refraction through the front correcting plate (lens). Using this arrangement, large photographic telescopes of extremely short focal length are practical (*see* Fig. 126). No eyepiece is used, and the image falls on a photographic plate shaped to the curved focal plane.

(*b*) A Maksutov-Gregorian (meniscus) telescope incorporates refractor and reflector principles. This type of instrument provides for large *f*/ratios within a short tube length. It can be used either photographically or visually and is suitable for compact, high-performance, portable telescopes. The telescope in Fig. 138 is constructed on similar principles.

Astronomical Photography

Very few visual observations are made with the large reflecting telescope in professional observatories. In modern times the photographic emulsion has completely superseded the human eye as a recorder of astronomical information. Photography has many advantages over the direct visual method:

(i) The record is permanent.

(ii) The photographic plate can be inspected, measured and analysed at leisure.

(iii) By time-exposure methods it can record faint objects far beyond the limit of direct vision.

(iv) Wavelengths can be photographed which are not visible to the human eye, particularly with colour emulsions.

In some instruments it would not be possible to observe visually, for they are constructed primarily as telescopic cameras, and the photographic plate is specially manufactured to a curved surface to suit the characteristics of the focal plane, such as in the Schmidt telescope (Fig. 172a).

Electronic Image Tubes

Although the use of the photographic plate has greatly increased the efficiency of the ordinary optical telescope as a recorder of astronomical information, the system is still very inefficient in respect to the received radiation. Even the most sensitive photographic emulsion requires at least 1000 incident light photons to be sure of recording one single 'bit' of permanent information.

As long ago as 1937, the French physicist Lallemand reasoned that by exploiting the photo-electric effect, gains of upwards of 100 could be realized over ordinary methods. By 1951, Lallemand had perfected a device which converted the visual image collected by the ordinary telescope into an enhanced electronic image much more intense than the visual one. The device had the effect of light amplification, and by 1956 it had reached a stage where, when used with a 47-in telescope, it could perform the equivalent task of a 275-in light collector.

Although there are still some practical difficulties to overcome, Lallemand's image tubes, and others developed elsewhere, have already greatly extended the range of present-day optical telescopes. By the 1980s it is expected that image tubes will enable optical telescopes to operate at distances at present restricted to radio telescopes.

The Spectroscope

The spectroscope is an instrument which was developed in the mid-nineteenth century as a result of the earlier chromatic light studies of Newton, Wollaston and Fraunhofer (p. 37) and is now one of the most powerful tools of astrophysical research.

All early spectroscopes employed glass prisms either singly or in combination to disperse the incident light into its visible chromatic wavelengths, but nowadays *diffraction* gratings are preferred since they provide a more even spread. These can take the form of *transmission* or *reflecting* gratings and consist of upwards of 15,000 machine-ruled grooves per inch on a polished glass surface or a metallic mirror.

The term spectroscope is restricted to the version used for direct visual purposes. In photographic work it is termed a *spectrograph*, and many large telescopes have one permanently attached. A *spectroheliograph* and a *spectrohelioscope* are instruments specially developed for work on the Sun.

Radio Telescopes

The great radio telescopes of today are direct descendants of the primitive radio telescope* devised by Karl Jansky in 1932 to detect the first discrete radio signals from outer space. Shortly after, Grote Reber, the first radio amateur astronomer, built a steerable parabolic reflector 30 ft in diameter in order to investigate Jansky's earlier sources. Reber found more, and his results were published early in World War II, but attention was then directed elsewhere. However, towards the end of the War, radio experiments, using standard military equipment, focused attention on the future possibilities of radio astronomy, which today has blossomed into a completely new science. Radio telescopes are the most important tools at our disposal in exploring the far corners of the Universe which lie at distances at present out of reach of the largest optical telescopes.

The advantage of radio telescopes over optical ones is that they can receive wavelengths from the radio frequency window between 1 cm and 20 m which are able to penetrate the atmosphere (Fig. 42). Wavelengths less than 1 cm are absorbed by the atmosphere, and those above 20 m are reflected back into space.

Among other advantages is that through radio waves we are able to explore *behind* the absorbing dust clouds of space which are opaque to radiation received by optical telescopes. Observation can also be made throughout the 24-hour period, for the brightly lit sunlight sky has no effect on incoming radio wavelength frequencies.

Radio telescopes take two principal forms of construction:

(i) Parabolic dishes consisting of a metal mesh or solid metallic surface.
(ii) Linear arrays of aerials arranged in various patterns.

Parabolic reflectors can be fixed or steerable structures. Unlike the optical reflecting telescopes their surfaces do not require great accuracy in figuring. They can be of sheet metal or wire mesh formed to an accuracy of about one-tenth the radio wavelength that has to be studied. The shape of the parabolic reflector is very reminiscent of that familiar domestic appliance, the electric fire, and has been aptly termed the 'dish'. In a similar manner to the optical reflector the radio 'dish' collects electromagnetic radiation and focuses the signals to a radio receiver where they are amplified and then fed to various sophisticated recording devices in a nearby control room (Fig. 129).

Also like optical telescopes the aperture determines the resolving power and the telescope's ability to penetrate deep into space. The resolving power of the 200-in Mt Palomar telescope is about one-tenth second of arc, while

* Which incorporated part of a Ford model T motor-car for some of the mounting.

the 250-ft radio telescope at Jodrell Bank (Fig. 163) has a resolution of 1°. Therefore it can be seen that radio telescopes require very large apertures in order to achieve the fine resolution obtained by the traditional optical telescopes.

The size limitation of steerable parabolic reflectors is one of construction and working stability rather than a technical one. The Jodrell Bank 250-ft dish is the largest operational steerable paraboloid in the northern hemisphere. In Australia the Parkes 210-ft radio telescope (Fig. 128) is of a similar design. *Fixed parabolic* reflectors can overcome some of the constructional problems, but since the area of sky to be studied must rotate towards the telescope, they are not as convenient to use.

The *linear radio telescope arrays* are arranged so that each aerial conducts its own 'bit' of received signal to a central receiver. This type of telescope has marked advantages in increasing the resolving power and can be used as an *interferometer* which can measure actual star diameters. The most efficient system of all is the use of aerials which form a cross, such as the Mills Cross radio telescope constructed for Sydney University (Fig. 165a, b), which literally provides a much sharper 'picture' of a radio source.

Other types of arrangement can also be made by using both parabolic and linear arrays to create aperture synthesis where only two small sections of a designed aperture need actually be constructed in order to gain the resolution characteristics of a much larger diameter radio telescope.

Two separate radio telescopes can be used for aperture synthesis. One of the longest link-ups of radio telescopes is between the Crimean-Astrophysical Observatory in the Soviet Union and the Haystack Observatory in Mass., U.S.A. – a distance of 7350 km. At present the length of baseline between two telescopes in such a system is only limited by the size of the Earth. However, in future it is likely that a radio telescope will be sited on the Moon so that the baseline can be extended to over 380,000 km.

Radio telescopes have many applications to astronomy. Apart from the investigation of objects beyond range of optical telescopes, they are also used in planetary studies, and in meteor astronomy they have made dramatic discoveries of daytime summer meteor showers which were previously unknown. Radio telescopes were also successful in detecting methyl cyanide (CH_3CN) in Comet Kohoutek during its apparition in 1973–4. Methyl cyanide is known to be present in interstellar clouds, and this provided the first direct evidence that comets contain interstellar material in their make-up.

APPENDIX 1

Amateur Astronomy

Astronomy is one of the most fascinating hobbies which can be pursued at all levels of interest. One can be a casual stargazer and simply observe the seasonal shifts of the star sphere and watch 'old friends' reappear and then disappear again for a short time as the year progresses; or one can aim towards becoming a world-famous dedicated amateur, such as was Sir William Herschel, and nightly sweep the heavens to make the discovery of a new comet or exploding star. If observation of the heavens does not appeal, or if one lives in a city where the stars are hidden by twentieth-century smog, one can be an armchair-astronomer and reflect on the different cosmologies of the Universe.

Although a great deal of stargazing can be done by simple naked-eye observation, a pair of inexpensive binoculars or a small astronomical telescope greatly extends the boundaries of celestial exploration. For those thus equipped, selected lists of suitable objects for observation can be found in the appendices. All these objects are marked by symbols on the star charts provided. To locate any object, look up its constellation name or its RA and Dec (p. 34) and select the appropriate map. If you prefer to work directly off the sky maps, check the symbols with the legend to identify any object shown and refer back to the selected list.

When first going outdoors, allow time (about 5 minutes) for the eyes to become fully dark-adapted and don't use a bright light to locate an object on the sky chart. Rather use a hand torch and cover it with red coloured paper or cloth, since red light does not interfere with one's night-vision capacity. To locate faint objects, using binoculars, choose a really dark, cloudless night when the Moon is absent. The atmosphere is particularly transparent after a rainstorm, and objects which are at other times faint stand out boldly.

In cold weather wrap up well and don't stand for long periods on concrete or hard pavement, for this will transmit the heat away from the feet. Use a piece of board.

If you become truly bitten by the astronomy bug, there are thousands of amateur societies scattered all over the world ready to enrol you as a member. Life-long friendships can be made, and the camaraderie among amateur astronomers is truly wonderful. Finally, amateur astronomers compose *both sexes*. Many famous amateur astronomers have been women, and some like Caroline Herschel have made important discoveries.

APPENDIX 2

THE BRIGHTEST STARS

Star	Apparent Magnitude	Absolute Magnitude	Spectral Type	Distance (Light Years)
Sirius (α Canis Majoris)	−1·4	+1·4	A1	8·7
Canopus (α Carinae)	−0·7	−4·4	F0	181·0
α Centauri	−0·3	+4·2	G2, K1	4·23
Arcturus (α Bootis)	−0·1	−0·2	K1	35·86
Vega (α Lyrae)	0·0	+0·5	A0	26·4
Capella (α Aurigae)	0·1	−0·6	G8, G0	45·64
Rigel (β Orionis)	0·1	−7·0	B8	880·0
Procyon (α Canis Minoris)	0·4	+2·7	F5	11·4
Achernar (α Eridani)	0·5	−2·2	B5	114·1
β Centauri	0·6	−5·0	B1	423·8
Betelgeuse (α Orionis)	0·4	−5·9	M2	586·0
Altair (α Aquilae)	0·8	+2·3	A7	16·4
Aldebaran (α Tauri)	0·8	−0·8	K5	68·46
β Crucis	0·8	−3·7	B1, B3	260·8
Antares (α Scorpii)	0·9	−4·7	M1, B	423·8
Spica (α Virginis)	1·0	−3·1	B1	211·9
Pollux (β Geminorum)	1·2	+1·0	K0	34·8
Fomalhaut (α Piscis Austrini)	1·2	+1·9	A3	22·82
Deneb (α Cygni)	1·3	−7·2	A2	1,630·0
α Crucis	1·3	−4·3	B0	423·8

APPENDIX 3

THE NEAREST STARS

Star	Apparent Magnitude	Absolute Magnitude	Spectral Type	Distance (Light Years)
Proxima Centauri	10·7	15·1	M5	4·2
α Centauri A	0·0	4·3	G2	4·2
α Centauri B	1·4	5·7	K4	4·3
Barnard's Star	9·5	13·2	M5	5·9
Wolf 359	13·7	16·8	M6	7·6
Lalande 21185	7·5	10·4	M2	8·1
Sirius A	−1·4	1·4	A1	8·7
Sirius B	8·7	11·5	w A5	8·7
Luyten 726–8 A	12·4	15·2	M6	8·7
Ross 154	10·6	13·3	M5	9·3
Ross 248	12·3	14·7	M6	10·3
ε Eridani	3·7	6·1	K2	10·7
Ross 128	11·1	13·5	M5	10·9
Luyten 789–6	12·6	14·9	M6	11·0
61 Cygni A	5·2	7·5	K5	11·2
61 Cygni B	6·0	8·3	K7	11·2
Procyon A	0·4	2·7	F5	11·4
Procyon B	10·8	13·1	w F	11·4
ε Indi	4·7	7·0	K5	11·4
Struve 2398 A	8·9	11·1	M4	11·5

w = white dwarf

APPENDIX 4

CONSTELLATIONS

Constellation	Genitive ending	Meaning	Abbreviations
Andromeda	-dae	Chained Maiden	And
Antlia	-liae	Air Pump	Ant
Apus	-podis	Bird of Paradise	Aps
Aquarius	-rii	Water Bearer	Aqr
Aquila	-lae	Eagle	Aql
Ara	-rae	Altar	Ara
Aries	-ietis	Ram	Ari
Auriga	-gae	Charioteer	Aur
Bootes	-tis	Herdsman	Boo
Caelum	-aeli	Chisel	Cae
Camelopardus	-di	Giraffe	Cam
Cancer	-cri	Crab	Cnc
Canes Venatici	-num -corum	Hunting Dogs	CVn
Canis Major	-is -ris	Great Dog	CMa
Canis Minor	-is -ris	Small Dog	CMi
Capricornus	-ni	Sea Goat	Cap
Carina	-nae	Keel of the Ship	Car
Cassiopeia	-peiae	Lady in Chair	Cas
Centaurus	-ri	Centaur	Cen
Cepheus	-phei	King	Cep
Cetus	-ti	Whale	Cet
Chamealeon	-ntis	Chameleon	Cha
Circinus	-ni	Compasses	Cir
Columba	-bae	Dove	Col
Coma Berenices	-mae -cis	Berenice's Hair	Com
Corona Australis	-nae -lis	Southern Crown	CrA
Corona Borealis	-nae -lis	Northern Crown	CrB
Corvus	-vi	Crow	Crv
Crater	-eris	Cup	Crt
Crux	-ucis	Southern Cross	Cru
Cygnus	-gni	Swan	Cyg
Delphinus	-ni	Dolphin	Del
Dorado	-dus	Swordfish	Dor
Draco	-onis	Dragon	Dra
Equuleus	-lei	Little Horse	Equ
Eridanus	-ni	River Eridanus	Eri
Fornax	-acis	Furnace	For
Gemini	-norum	Heavenly Twins	Gem
Grus	-ruis	Crane	Gru
Hercules	-lis	Kneeling Giant	Her
Horologium	-gii	Clock	Hor
Hydra	-drae	Water Monster	Hya
Hydrus	-dri	Sea-serpent	Hyi
Indus	-di	Indian	Ind
Lacerta	-tae	Lizard	Lac
Leo	-onis	Lion	Leo
Leo Minor	-onis -ris	Small Lion	LMi
Lepus	-poris	Hare	Lep
Libra	-rae	Scales	Lib
Lupus	-pi	Wolf	Lup
Lynx	-ncis	Lynx	Lyn
Lyra	-rae	Lyre	Lyr

Constellation	Genitive ending	Meaning	Abbreviations
Mensa	-sae	Table (Mountain)	Men
Microscopium	-pii	Microscope	Mic
Monoceros	-rotis	Unicorn	Mon
Musca	-cae	Fly	Mus
Norma	-mae	Square	Nor
Octans	-ntis	Octant	Oct
Ophiuchus	-chi	Serpent Bearer	Oph
Orion	-nis	Hunter	Ori
Pavo	-vonis	Peacock	Pav
Pegasus	-si	Winged Horse	Peg
Perseus	-sei	Perseus	Per
Phoenix	-nicis	Phoenix	Phe
Pictor	-ris	Painter's Easel	Pic
Pisces	-cium	Fishes	Psc
Piscis Austrinus	-is -ni	Southern Fish	PsA
Puppis	-ppis	Poop (Stern of Ship)	Pup
Pyxis (=Malus)	-xidis	Compass	Pyx
Reticulum	-li	Net	Ret
Sagitta	-tae	Arrow	Sge
Sagittarius	-rii	Archer	Sgr
Scorpius	-pii	Scorpion	Sco
Sculptor	-ris	Sculptor	Scl
Scutum	u-ti	Shield	Sct
Serpens (Caput and Cauda)	-ntis	Serpent (Head and Tail)	Ser
Sextans	-ntis	Sextant	Sex
Taurus	-ri	Bull	Tau
Telescopium	-pii	Telescope	Tel
Triangulum	-li	Triangle	Tri
Triangulum Australe	-li -lis	Southern Triangle	TrA
Tucana	-nae	Toucan	Tuc
Ursa Major	-sae -ris	Great Bear	UMa
Ursa Minor	-sae -ris	Little Bear	UMi
Vela	-lorum	Sails	Vel
Virgo	-ginis	Virgin	Vir
Volans	-ntis	Flying Fish	Vol
Vulpecula	-lae	Little Fox	Vul

For the history and mythology of each constellation and a comprehensive star-by-star guide to the constellations, see the author's *What Star Is That?* Viking Press, New York; Thames and Hudson, London, 1971.
For a more exhaustive treatment of comets, meteorites and meteoroids, see the author's *Comets, Meteorites and Men*, Robert Hale, London 1973; Taplinger, New York, 1974.

APPENDIX 5

Star Nomenclature

The principal stars in each constellation are designated by the letters of the Greek alphabet or by their Arabic names:

α	Alpha	ι	Iota	ϱ	Rho
β	Beta	κ	Kappa	σ	Sigma
γ	Gamma	λ	Lambda	τ	Tau
δ	Delta	μ	Mu	υ	Upsilon
ε	Epsilon	ν	Nu	φ	Phi
ζ	Zeta	ξ	Xi	χ	Chi
η	Eta	o	Omicron	ψ	Psi
θ	Theta	π	Pi	ω	Omega

Usually, but not always, the brightest star is designated α, and the next brightest β and so on. When Greek letters are exhausted for a particular constellation, Roman letters are used. Flamsteed, the first Astronomer Royal, introduced a number system 1, 2, 3, etc. for stars in each constellation (in their order of Right Ascension) so that, for example, the brightest star Aldebaran is α Tauri, or 87 (Flamsteed); it could also be identified by a designation contained in a specialist catalogue (*see below*) or by its RA and Dec, but usually it is known simply by its Greek designation or its popular Arabic name.

Examples of Specialist Star Catalogues

Aitken's Double Star Catalogue (A.D.C.)
Astrographic Catalogue (A.Z.)
British Association Catalogue (B.A.C.)
Bonn Durchmusterung (B.D.)
Boss' General Catalogue (G.C.)
Burnham's General Catalogue of Double Stars (B.G.C.)
Lacaille's Catalogue of Southern Stars (Lac.)
Sir William Herschel's Catalogue (H)
Sir John Herschel's Catalogue (h)
Sir John Herschel's Catalogue of Sir William Herschel's Double Stars (Hh)
F. G. W. Struve's list (Σ)
Otto Struve's list ($O\Sigma$)
Pulkova Catalogue ($O\Sigma\Sigma$)
Contractions are shown in brackets.

Variable stars, except for those with Greek designations, are denoted by a Roman capital letter beginning with R to Z and preceded by the name of the constellation, e.g. R Andromedae. When more than nine variables were discovered in one constellation, double letters were introduced, e.g. RR, RZ, etc. to ZZ. Later, with further discoveries, more combinations had to be introduced, so it was continued with AA, AZ, etc. In this method there are 325 combinations available. Nowadays, since faint variable stars are discovered in large numbers, any new variable found in a constellation which has already 334 variables, is designated by an Arabic number only, and is preceded with a Roman capital V, e.g. V 515 Persei. The best known catalogue of variable stars is *Catalogue of Variable Stars* by Kukarkin and Parengo.

The brightest galaxies, nebulae and star clusters were first catalogued by the Frenchman Charles Messier in 1784, and his original list numbering 103 objects, M 1, M 2, etc., is still in current use today. However, the discoveries of William Herschel and others soon increased the total number into many thousands. In 1888, the British astronomer Dreyer formulated the *New General Catalogue* or the NGC. For example the Great Nebula in Andromeda is M 31 or NGC 224. A supplementary *Index Catalogue* or the IC followed at a later date.

APPENDIX 6

TABLE OF LIMITING MAGNITUDES AND
TELESCOPIC RESOLVING POWER
(*see also* p. 214)

The ability to see faint stars, or the limiting magnitude, depends principally on the aperture of the binoculars or the telescope used. Also the ability to resolve fine planetary detail, or split double stars, depends on aperture. But it must be remembered that inferior instruments or poor observing conditions may lead to inferior performance. Low-power binoculars in the range of $\times 6 - \times 12$ generally have insufficient magnification to split double stars less than $1'·5$ apart.

Diameter of Object Glass or Mirror	Naked eye	1-in	2-in	3-in	4-in	6-in	8-in	12-in
Closest Stars split	3'	4"·56	2"·28	1"·52	1"·14	0"·76	0"·57	0"·38
Magnitude Limit (approx.)	6	9·0	10·5	11·4	12·0	12·9	13·5	14·4

The limiting magnitude of diffuse objects such as galaxies, nebulae and faint comets depends on their size, atmospheric transparency and the amount of scattered light in the sky. Generally speaking, the limiting magnitude of galaxies, nebulae and faint comets is two magnitudes brighter than for stars (e.g. a 1-in telescope: mag 9 stars, but mag 7 diffuse objects). However, when the atmosphere is particularly transparent and the sky dark, their threshold magnitudes will approach those of stars.

247

APPENDIX 7

INTERESTING DOUBLE STARS

Star and Constellation	Magnitudes	Distance Apart (sec.)	Colours	Position (1950) RA	Dec
γ And	3·0, 5·0	10	yellow, blue	02h00m	+42°·1
ι Cnc	4·4, 6·5	30	yellow, blue	08 44	+29°·0
α CVn	3·2, 5·7	20	blue, blue	12 54	+38°·6
α¹, α² Cap	4·0, 3·8	376	yellow, yellow	20 15	−12°·7
ι Cas	4·2, 7·1, 8·1	2, 7	yellow, blue, blue	02 25	+67°·2
α Cen	0·3, 1·7	4	yellow, red	14 37	−60°·6
ς CrB	4·1, 5·0	6	white, blue	15 38	+36°·8
α Cru	1·4, 1·9	5	blue, blue	12 24	−63°·8
β Cyg	3·0, 5·3	35	yellow, blue	19 29	+27°·8
γ Del	4·0, 5·0	10	yellow, green*	20 44	+16°·0
ν Dra	4·6, 4·6	62	white, white	17 31	+55°·2
ψ Dra	4·0, 5·2	31	yellow, purple*	17 43	+72°·2
32 Eri	4·0, 6·0	7	yellow, blue	03 52	−03°·1
α Gem	2·7, 3·7	5	white, white	07 31	+32°·0
α Her	3·0, 6·1	4	orange, green*	17 12	+14°·4
ε Lyr	4·6, 4·9	208	yellow, blue	18 43	+39°·6
ε¹ Lyr	4·6, 6·3	3	yellow	18 43	+39°·6
ε² Lyr	4·9, 5·2	2	blue	18 43	+39°·6
β Ori	4·0, 10·3, 2·5, 6·3	(Quadruple)	blues	05 36	−02°·6
α Sco	1·2, 6·5	3	red, white	16 26	−26°·3
β Tuc	4·5, 4·5	26	blue, white	00 29	−63°·2
ς UMa	2·4, 4·0	14	white, white	13 22	+55°·2
α UMi	2·5, 8·8	19	yellow, blue	01 49	+89°·0
γ Vir	3·6, 3·7	6	white, yellow	12 39	−01°·2

Note: Refer to Appendix 6 for aperture required to split different stars.
* 'Dazzle tint', *see* p. 214.

APPENDIX 8

FAMOUS STAR CLUSTERS

Constellation	Object	Position RA	Dec	Type	Mag	Remarks
Auriga	M 38	05h25m	+35°·8	Open	7·4†	
Auriga	M 37	05 49	+32°·6	Open	6·2†	
Cancer	M 44	08 37	+20°·2	Open	3·7*	'Praesepe' or 'the Beehive'
Canes Venatici	M 3	13 40	+28°·6	Globular	6·4†	
Cassiopeia	M 103	01 30	+60°·4	Open	7·4†	
Centaurus	NGC 3766	11 34	−61°·3	Open	5·1†	
Centaurus	ω	13 24	−47°·0	Globular	3·7*	½° diameter
Crux	NGC 4755	12 51	−60°·1	Open	5·2†	
Cygnus	M 39	21 30	+48°·2	Open	5·2†	
Gemini	M 35	06 06	+24°·4	Open	5·3†	
Hercules	M 13	16 40	+36°·6	Globular	5·7†	'Great Cluster'
Pegasus	M 15	21 28	+12°·0	Globular	6·0†	
Perseus	NGC 869 and 884	02 18	+56°·9	Open	4·4* 4·7*	'Swordhandle Double Cluster'
Perseus	M 34	02 39	+42°·5	Open	5·5†	
Sagittarius	M 23	17 54	−19°·0	Open	6·9†	
Scorpius	M 6	17 37	−32°·2	Open	5·3†	
Scorpius	M 7	17 51	−34°·8	Open	3·2†	
Scutum	M 11	18 48	−06°·8	Open	6·3*	
Taurus	M 45	03 44	+24°·0	Open	1·6*	'the Pleiades'
Triangulum Australe	NGC 6025	15 59	−60°·4	Open	5·8†	
Tucana	NGC 104	00 22	−72°·4	Globular	3·0*	Star 47 Tucanae

* Visible to naked eye.   † Visible in binoculars.

APPENDIX 9

NAKED-EYE VARIABLE STARS

Variable		Position		Magnitudes		Period (days)	Type
		RA	Dec	Max.	Min.		
α	Cas	00h39m	+56° 16′	2·5	3·1	—	Irr
γ	Cas	00 54	+60 27	1·6	3·0	—	Irr
ο	Cet†	02 17	−03 12	2·0	10·1	331·48	LP
β	Per	03 05	+40 46	2·2	3·5	2·8673	EA
λ	Tau	03 58	+12 21	3·5	4·0	3·9530	EA
ε	Aur	04 58	+43 45	3·7	4·5	9883	EA
AE	Aur*	05 13	+34 15	5·4	6·1	—	Irr
α	Ori	05 53	+07 24	0·4	1·3	2070	SR
η	Gem	06 12	+22 31	3·1	3·9	233·4	SR
R	Hya†	13 27	−23 01	3·5	10·9	387	LP
δ	Lib*	14 58	−08 19	4·8	5·9	2·3273	EA
α	Sco	16 26	−26 19	1·2	1·8	1733	SR
g	Her*	16 27	+41 59	4·6	6·0	80	SR
μ¹	Sco	16 48	−37 58	3·0	3·3	1·4463	β Lyr
α¹	Her	17 12	+14 27	3·0	4·0	100	SR
X	Sgr*	17 44	−27 49	5·0	6·1	7·0122	δ Cep
β	Lyr	18 48	+33 18	3·4	4·3	12·9080	β Lyr
R	Lyr	18 54	+43 53	4·0	5·0	50	SR
χ	Cyg†	19 49	+32 47	2·3	14·3	406·66	LP
P	Cyg	20 16	+37 53	3·0(?)	6·0(?)	—	N
μ	Cep	21 42	+58 33	3·6	5·1	—	SR
δ	Cep	22 27	+58 10	3·9	5·0	5·3663	δ Cep
ϱ	Cas	23 52	+57 13	4·1	6·2	—	Irr
R	Cas†	23 56	+57 07	4·8	13·6	430·93	LP

Note: **SR.** Semi-Regular star; **EA.** Eclipsing Binary star of Algol-type; δ Cep. Delta Cepheid-type variable; β Lyr. Beta Lyrae-type variable; Irr. Irregular period star; LP. Long-period star; N. Novae-type star. * Difficult naked-eye object. † Naked-eye only at maximum.

APPENDIX 10

INTERESTING GALAXIES AND NEBULAE

Constellation	Object	Position		Type	Mag	Remarks
		RA	Dec			
Andromeda	M 31	00h 40m	+41°·0	Spiral Gal.	4·8*	'Great Nebula'
Canes Venatici	M 51	13 28	+47°·4	Spiral Gal.	8·1	'Whirlpool Nebula'
Dorado	NGC 2070	05 39	−69°·2	Diffuse Neb.	—*	Visible to naked eye
Lyra	M 57	18 52	+33°·0	Planetary Neb.	9·3‡	'Ring Nebula'
Orion	M 42	05 33	−05°·4	Diffuse Neb.	4·0*	'Great Nebula'
Sagittarius	M 20	17 59	−23°·0	Diffuse Neb.	6·9†	'Trifid Nebula'
Sagittarius	M 8	18 01	−24°·4	Diffuse Neb.	6·8†	'Lagoon Nebula'
Sagittarius	M 17	18 18	−16°·2	Diffuse Neb.	7·0†	'Omega Nebula'
Taurus	M 1	05 32	+22°·0	Diffuse Neb.	8·4‡	'Crab Nebula'
Triangulum	M 33	01 31	+30°·4	Spiral Gal.	6·7	Large and faint
Ursa Major	M 81	09 52	+69°·3	Spiral Gal.	7·9‡	Connected to M 82
Ursa Major	M 97	11 12	+55°·3	Planetary Neb.	11·0**	'Owl Nebula'
Vulpecula	M 27	19 58	+22°·6	Planetary Neb.	7·6‡	'Dumb Bell Nebula'

* Visible to naked eye.   † Visible in binoculars.   ‡ Requires 2-in telescope.
** Requires 3-in telescope.

APPENDIX II

SUN, MOON AND PLANETS (PHYSICAL ELEMENTS)

Planet etc.	Equatorial Diameter (km)	Sidereal Period of Axial Rotation	Inclination °	Density Water =1	Escape Velocity km s⁻¹	Equatorial Diameter Earth=1	Surface Gravity	Mean Visual Mag	Albedo
Sun	1,392,000	25d·380	7 15	1·409	617·5	109·12	27·90	−26·8	—
Moon	3,476	27·322	1 32	3·350	2·38	0·2725	0·1653	−12·7	0·07
Mercury	4,840	58·7	0	5·50	4·27	0·382	0·381	−1·8	0·06
Venus	12,390	243·0	177 (3°) 27	5·25	10·36	0·9489	0·9032	−4·4	0·75
Earth	12,760	23h56m04s	23	5·517	11·18	1·0000	1·0000	—	0·36
Mars	6,800	24 37 23	23 59	3·94	5·03	0·5320	0·3799	−2·8	0·16
Jupiter	142,800	9 50 30	3 04	1·330	60·22	11·197	2·643	−2·2	0·73
Saturn	119,400	10 14	26 44	0·706	36·25	9·355	1·159	+0·7	0·76
Uranus	47,600	10 49	97 53	1·60	22·4	3·70	1·11	+5·8	0·93
Neptune	48,400	15 30	29 30	1·77	23·9	3·79	1·21	+7·8	0·84
Pluto	5,900	6d·390	?	5·5	5·1	0·47	0·47	+14·5	0·14

APPENDIX 11a

THE PLANETS (ORBITAL ELEMENTS)

Planet	Mean Distance astronomical units	Mean Distance millions of kilometres	Perihelion Distance (A.U.)	Aphelion Distance (A.U.)	Sidereal Period (d)	Mean Synodic Period (d)	Inclination to the Ecliptic i	Eccentricity	Mean Orbital Velocity km sec⁻¹
Mercury	0.3870987	57.91	0.3075	0.4667	87.969	115.88	7° 00′ 15.1″	0.2056287	47.87
Venus	0.7233322	108.29	0.7184	0.7282	224.701	583.92	3 23 39.6	0.0067868	35.02
Earth	1.0000000	149.60	0.9833	1.0167	365.256	—	—	0.0167213	29.79
Mars	1.5236915	227.94	1.381	1.666	686.980	779.94	1 50 59.5	0.0933782	24.13
Jupiter	5.2028039	778.34	4.951	5.455	4332.59	398.88	1 18 17.1	0.0484533	13.06
Saturn	9.5388437	1,427.7	9.008	10.070	10759.20	378.09	2 29 21.9	0.0556436	9.65
Uranus	19.181876	2,869.6	18.28	20.09	30685.0	369.66	0 46 23.3	0.0472394	6.80
Neptune	30.057912	4,496.7	29.80	30.32	60190.2	367.49	1 46 21.8	0.0085832	5.43
Pluto	39.5	5,900.0	29.7	39.5	90700.0	366.74	17 10	0.247	4.74

APPENDIX 12

SATELLITES OF THE SOLAR SYSTEM

Planet and Satellite	Distance from planet (km)	Mean Sidereal Period (d)	Mean Synodic Period d h m s	Inclination °	Eccentricity	Diameter (km)	Mean Visual Mag
Earth Moon	384,000	27.321661	29 12 44 02.9	23.4	0.0549	3,476	−12.7
Mars							
I Phobos	9,300	0.318910	7 39 26.6	1.1	0.0210	22	11.6
II Deimos	23,500	1.262441	1 06 21 15.7	1.8	0.0028	6	12.8

SATELLITES OF THE SOLAR SYSTEM—continued

Satellite	Distance (km)	Period (days)	Period (d h m s)	Inclination	Eccentricity	Diameter	Mag.
Jupiter							
V	181,000	0·498179	11 57 27·6	0·4	0·003	200	13·0
I Io	422,000	1·769138	1 18 28 35·9	0·0	0·000	3,900	4·8
II Europa	671,000	3·551181	3 13 17 53·7	0·5	0·0001	2,900	5·2
III Ganymede	1,070,000	7·154553	7 03 59 35·9	0·2	0·0014	5,000	4·5
IV Callisto	1,883,000	16·689018	16 18 05 06·9	0·2	0·0074	4,500	5·5
VI	11,470,000	250·5662	265 22 43	24	0·1580	100	13·7
X	11,850,000	259·2188	275 17 09	27	0·1074	20	18·6
VII	11,740,000	259·6528	276 04 56	28	0·2072	30	16·0
XII	21,200,000	631	551	147	0·169	30	18·8
XI	22,560,000	692	597	163	0·207	20	18·1
VIII	23,500,000	744	635	148	0·410	20	18·8
IX	23,700,000	758	645	157	0·275	20	18·3
Saturn							
X Janus	150,000	0·7490	17 59	0·0	0·0	300	14·0
I Mimas	186,000	0·942422	22 37 12·4	1·5	0·0202	500	12·1
II Enceladus	238,000	1·370218	1 08 53 21·9	0·0	0·0045	600	11·8
III Tethys	295,000	1·887803	1 21 18 54·8	1·1	0·000	1,000	10·3
IV Dione	377,000	2·736916	2 17 42 09·7	0·0	0·0022	1,000	10·4
V Rhea	527,000	4·517503	4 12 27 56·2	0·3	0·0010	1,300	9·8
VI Titan	1,222,000	15·945448	15 23 15 31·5	0·3	0·0292	5,000	8·4
VII Hyperion	1,481,000	21·276657	21 07 39 05·7	0·6	0·1042	500	14·2
VIII Iapetus	3,560,000	79·33085	79 22 04 59	14·7	0·0283	1,100	11·0
IX Phoebe	12,950,000	550·337	523 13	150	0·1633	200	16·5
Uranus							
V Miranda	128,000	1·413499	1 09 55 31	0·0	0·00	300	16·5
I Ariel	192,000	2·520384	2 12 29 39·0	0·0	0·0028	800	14·4
II Umbriel	267,000	4·144183	4 03 28 25·8	0·0	0·0035	600	15·3
III Titania	438,000	8·705876	8 17 00 01·2	0·0	0·0024	1,000	14·0
IV Oberon	586,000	13·463262	13 11 15 36·5	0·0	0·0007	1,000	14·2
Neptune							
I Triton	353,000	5·876844	5 21 03 29·8	159·9	0·000	3,700	13·5
II Nereid	5,600,000	359·881	362 01	27·7	0·7493	300	18·7

ASTRONOMICAL SIGNS AND SYMBOLS

Solar System

- ⊙ Sun
- ☿ Mercury
- ♀ Venus
- ⊕ Earth
- ♂ Mars
- ♃ Jupiter
- ♄ Saturn
- ♅ Uranus
- ♆ Neptune
- ♇ Pluto
- ☽ Moon
- ● New Moon
- ○ Full Moon
- ☽ First Quarter
- ☾ Last Quarter
- ⑤ Minor planet (number in a circle)
- ☄ Comet
- ♈ Vernal Equinox
- ☌ Conjunction
- □ Quadrature
- ☍ Opposition
- ☊ Ascending Node
- ☋ Descending Node

Signs of the Zodiac (Fig. 81)

Angular Measure

degrees (°)
minutes (′)
seconds (″)

Astrophysics

m (or mag)	Apparent Magnitude
M	Absolute Magnitude
m_v	Apparent Visual Magnitude
m_{pg}	Apparent Photographic Magnitude
λ	Wavelength
Å	Ångstrom Unit $= 10^{-7}$ mm
μ	Micron $= 10^{-3}$ mm

ASTRONOMICAL UNITS

	Miles (mi)	Kilometres (km)	Astronomical Units (A.U.)	Light Years (l.y.)	Parsecs (psc)
	1	1·6093	$1·076 \times 10^{-8}$	$1·701 \times 10^{-13}$	$0·522 \times 10^{-13}$
	0·62137	1	$0·669 \times 10^{-8}$	$1·057 \times^{-13}$	$0·324 \times 10^{-13}$
1 A.U.	$9·29 \times 10^7$	$1·495 \times 10^8$	1	$0·158 \times 10^{-5}$	$0·485 \times 10^{-5}$
1 l.y.	$5·88 \times 10^{12}$	$9·460 \times 10^{12}$	63,280	1	0·3068
1 psc	$1·916 \times 10^{13}$	$3·084 \times 10^{13}$	206,265	3·260	1

Megaparsec (mpsc) $= 10^6$ psc   Kiloparsec kpsc $= 10^3$ psc
Velocity of light $= 6·70 \times 10^8$ (mi/hr) $= 186,000$ (mi/sec) $= 299,791$ (km/sec)

Temperature Conversion

Units	To	
Fahrenheit (°F)	°C	Subtract 32 and multiply by 5/9
Centigrade (°C)	°F	Multiply by 9/5 and add 32
°C	Kelvin (°K)	Add 273·155

GLOSSARY

Absorption Lines (Fraunhofer Lines)
Dark lines appearing in a spectrum due to the absorption of each particular wavelength of light by a chemically identical substance, but from a cooler source.

Albedo
The percentage of sunlight reflected from a (planet) surface compared with the total amount received from the Sun.

Anomalistic Year
The time required for the Earth to revolve round the Sun in relation to its perihelion = 365·2596 days. Since the line of the apsides (q.v.) advances in the direction of the Earth's orbital motion, the *anomalistic year* = 0·0174 days *longer* than the *tropical year* (q.v.).

Apogee
The point in the Moon's (or artificial satellite's) orbit which is *farthest* from the Earth. Sometimes also used in relation to the orbit of the Earth round the Sun.

Apsides
The two points connecting a line across the major axis of an elliptical orbit (*see* p. 15).

Cassegrain (System)
A form of reflecting telescope in which the secondary mirror is convex, and the reflected light rays pass through an aperture in the primary mirror.

Continuous Spectrum
The unbroken spectrum of colours visible in the optical band of the electromagnetic spectrum extending between the infrared to the ultraviolet regions.

Cosmic Rays
Highly energized radiation falling on the Earth from sources in outer space. They consist chiefly of protons, electrons and alpha particles (helium nuclei), but they also contain much heavier atomic nuclei. Collisions with the atmospheric gases give rise to secondary interactions, and *cosmic ray showers* result.

Coudé (System)
An optical arrangement in telescopes where the light path is arranged to be reflected through the axis of the mounting to a fixed position. It is used to provide very long focal lengths which is necessary for high resolution spectroscopy.

Eclipse Year
The time between successive returns of the Sun to the Moon's ascending node (q.v.) = 346·62 days. So named because it represents the interval between eclipses which can only occur near a node (*see* p. 60).

Emission Lines
The bright lines appearing in a spectrum. They occur when an orbital electron in a gas falls back to a less energetic inner orbit and thus emits the excess energy as light.

Fluorescence
A property of absorbing light in one particular wavelength and emitting it in another wavelength. Unlike *phosphorescence* the phenomenon stops when the source of radiation is cut off. The gases in the coma of a comet's head shine due to fluorescence.

Gamma-Rays
Electromagnetic radiation given off by radioactive atoms as they decay.

Golden Number (*see* Metonic Cycle p. 52)
The number denoting the position of any year in the lunar (Metonic) 19-year cycle. The *golden number* is found by adding 1 to the given year and then dividing by 19; the *remainder* then = the *golden number* unless the remainder is = to 0 (zero) in which case the *golden number* is 19 (e.g. the *golden number* of 1972 = 16).

Gregorian Year
The mean length of year according to the Gregorian calendar = 365·243 *mean solar days*.

Ionization
A process by which an atom of matter loses one or more orbital electrons which causes it to become positively charged. The affixed symbol + = one electron lost, + + = two electrons lost, etc. Matter forming stars and nebulae is often strongly ionized.

Julian Year
The mean length of year according to the Julian calendar = 365·25 *mean solar days*.

Maksutov Telescope
A variety of telescope developed in the early 1940s chiefly by D.D. Maksutov, a Soviet astronomer. It utilizes both a front refracting lens (meniscus) and a spherical mirror, a combination often referred to as a *catadioptric system* similar to the Schmidt telescope (q.v.), but it has the advantage that it can also be used as a visual instrument.

Mean Solar Day
Because of the slight eccentricity of the Earth's orbit round the Sun, the length of the day as measured by the Sun varies throughout the year. The average value is called the *mean solar day* = 24^h 03^m $56^s.555$.

Mesons
Atomic particles with masses between the electron and proton. Mesons are found in *cosmic ray showers* and are extremely short-lived phenomena.

Perigee
The point in the Moon's (or artificial satellite's) orbit which is *nearest* to the Earth. Sometimes also used in relation to the orbit of the Earth round the Sun.

Photo-Electric Effect
An event occurring when an electron is ejected from its orbital position in an atom by a photon (q.v.). Electronic image tubes (image intensifiers) exploit this effect (*see* p. 239).

Photon
The smallest discrete unit, 'package' or *quantum* (q.v.) of light.

Plasma
A highly excited gas consisting almost entirely of ionized atoms or molecules; often referred to colloquially as *the fourth state of matter* (*see* also ionization).

Quantum
The smallest discrete and indivisible physical quantity of energy forms.

Schmidt Telescope
A variety of photographic telescope developed in the 1920s by B. Schmidt, a German amateur astronomer. It utilizes a front correcting lens (plate) and a spherical mirror. The Schmidt telescope, and its later development called a Super-Schmidt, can be constructed with very short focal lengths.

Sidereal Year
The period of the apparent revolution of the Sun in respect to the star sphere = the Earth's revolution in orbit round the Sun = 365·2564 *mean solar days* (q.v.).

Sungrazers
A colloquial name given to a group of comets of long period which at perihelion appear to graze the Sun's atmosphere. Sometimes the term is also applied to asteroids (minor planets) which have highly eccentric orbits allowing them to approach the Sun.

Tropical Year
The time required by the Sun to complete one revolution with respect to the vernal equinox = 365·242 *mean solar days*.

X-Rays
Electromagnetic radiation of the same kind as light, but of much shorter wavelength.

INDEX

Figures in bold face type are Fig. nos. Figures in italics refer to main page reference of a subject.